Coffee

Resources Series

Coffee

GAVIN FRIDELL

polity

First published in 2014 by Polity Press

Polity Press
65 Bridge Street
Cambridge CB2 1UR, UK

Polity Press
350 Main Street
Malden, MA 02148, USA

ISBN-13: 978-0-7456-7076-8
ISBN-13: 978-0-7456-7077-5(pb)

A catalogue record for this book is available from the British Library.

Typeset in 10.5 on 13pt Scala by
Servis Filmsetting Ltd, Stockport, Cheshire

For further information on Polity, visit our website: www.politybooks.com

For Sasha, Sebastian, and Kate

Contents

Abbreviations

ABIC	Brazilian Coffee Roasters' Association (Associação Brasileira da Indústria de Café)
ACPC	Association of Coffee-Producing Countries
BRICS	Brazil, Russia, India, China, and South Africa (group of states)
CAFE	Coffee and Farmer Equity program (Starbucks Coffee Company)
CAW 3000	Canadian Auto Workers, Local 3000
CLAC	Latin American and Caribbean Network of Small Fair Trade Producers (Coordinadora Latinoamericana y del Caribe de Pequeños Productores de Comercio Justo)
CSR	corporate social responsibility
DWSR	Dollar–Wall Street Regime
ECX	Ethiopian Commodity Exchange
EIPO	Ethiopian Intellectual Property Office
FAO	Food and Agriculture Organization of the United Nations
FLO	Fairtrade International
FNC	National Federation of Coffee Farmers (Federación Nacional de Cafeteros), Colombia
FOB	free on board
Fundeppo	Foundation of Organized Small Producers (Fundación de Pequeños Productores Organizados)
GLOBALG.A.P.	Global Good Agricultural Practices
IACA	Inter-American Coffee Agreement
IACO	Inter-African Coffee Organization
ICA	International Coffee Agreement

ICO	International Coffee Organization
IISD	International Institute for Trade and Development
NCA	National Coffee Association (United States of America)
NEZ	New Economic Zone (Vietnam)
ODI	Overseas Development Institute
PSD	Social Democratic Party (Partido Social Demócrata, Costa Rica)
SOE	state-owned enterprise
SPS	Small Producers' Symbol
TNC	transnational corporation
UN Comtrade	United Nations Commodity Trade Statistics
UNCTAD	United Nations Conference on Trade and Development
UNDP	United Nations Development Programme
VBARD	Vietnamese Bank of Agriculture and Rural Development
VBSP	Vietnam Bank for Social Policy
Vinacafe	Vietnam Coffee Corporation
WITS	World Integrated Trade Solution
WTO	World Trade Organization

Figures

Acknowledgments

The author would like to thank many colleagues and friends for their advice, inspiration, and support over the years, including Haroon Akram-Lodhi, Greg Albo, Bill Barrett, David Friesen, Ilan Kapoor, David McNally, Viviana Patroni, Darryl Reed, John Talbot, Steven Topik, and Tony Winson. Very special thanks are due Mark Gabbert, Martijn Konings, and Liisa North for our many lengthy chats related to coffee and beyond. An immeasurable debt is owed to all of those who agreed to take time from their busy lives to be interviewed over the years and facilitate my current and earlier research on coffee and commodities in Mexico, the Caribbean, Canada, and Europe. Christina Sayers and Amr El-Alfy were excellent graduate research assistants and Jenny Kaulback and Cassie MacDonald offered invaluable administrative support. Louise Knight and Pascal Porcheron from Polity Press, as well as the anonymous reviewers, provided superb guidance in improving the work and seeing it through. The author is indebted to the International Coffee Organization (ICO) for permitting him the unique opportunity to observe its annual meeting in London, England, in March 2013. Financial and institutional support from the Social Sciences and Humanities Research Council of Canada, the Canada Research Chair program, and Saint Mary's University is gratefully acknowledged.

Above all, I would like to thank my family, in particular my "espresso" club, Sebastian and Sasha, who have infused more energy into the house than any number of coffees could possibly accomplish, and Kate Ervine, as always my unwavering ally in life, without whose guidance and knowledge this book could not have been written. Any errors or omissions, of course, are entirely the author's own.

CHAPTER ONE

The global market and coffee statecraft

Following the global coffee market is a daunting task for any researcher, not least because of the dramatic ups and downs of coffee prices. When I began my graduate studies as a masters student in 1996, coffee prices were in the middle of a five-year recovery after a previous four-year collapse. The coffee composite indicator price, a commonly used estimate that combines different quality beans with different prices, dropped to extreme lows from 1990 to 1993, reaching as low as 54 cents per pound, only to recover starting in 1994, eventually reaching as high as $1.38 per pound. When I started my doctoral work in 1998, the mini-boom had already ended, with prices collapsing once again, this time to the lowest seen in 30 years and by some estimates the lowest prices in *real value* in over 100 years, taking into account historical rates of inflation. Prices dropped as low as 45 cents, causing a major global coffee crisis that left tens of thousands of farmers and rural workers confronting bankruptcy, migration, and hunger. Prices did not recover from the crisis years until 2007, when they crawled over $1 per pound and began to slowly increase. Then in 2011, as I began work on this book, prices boomed once again, reaching $1.95 per pound, the highest seen in decades, causing economic analysts to rush to celebrate the new "bull" market (see figure 1.1). As I wrapped up research for the book in the fall of 2013, prices had fallen, dropping to around $1, causing the International Coffee Organization (ICO) to raise concerns that "many producers may be selling

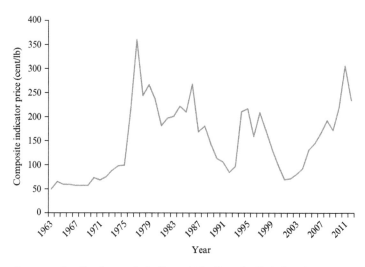

Source: 1976 coffee Composite Indicator Price from the United Nations Conference on Trade and Development (UNCTAD) statistical database (http://unctadstat.unctad.org), accessed July 30, 2013.

Note: Year 2000 = 100.

Figure 1.1 The coffee rollercoaster (taking into account inflation), 1963–2012.

at a price which is not remunerative compared to the cost of production." Prices then soared upward once more at the start of 2014, due to an unanticipated drop in global coffee supply caused by one of the worst droughts in decades in Brazil, the world's largest coffee producer. In the words of *Wall Street Journal* reporters, this reflected "how quickly global commodity markets can swing from concerns about oversupply to fears of a shortage."[1]

Charting the rollercoaster of global coffee prices, of course, is nothing compared to riding on it. Every dramatic drop and rise in the coffee rollercoaster brings with it immense social, economic, and political ramifications worldwide. More than just low prices, it is coffee's extreme price volatility that has major impacts on coffee growing, exporting, and import-ing states; coffee companies and consumers; and, of course, above all else, millions of coffee farmers and rural workers – there are approximately 25 million coffee farmer families in the world, over 70 percent of whom are small farmers owning less than 10 hectares of land. During boom times, coffee has a reputation for providing good incomes among farmers in Latin America, Africa, and Asia, where most of the world's coffee is grown due to the climatic needs of the tropical bean. Coffee is one of the most valuable commodities exported by the South, seconded only by oil and illegal drugs, and is capa-ble of generating immense wealth.

During frequent bust times, however, the face of coffee changes, poverty and hunger intensify, general despair emerges, and bankruptcies and unwanted migrations in search of work are the outcomes for countless thousands. Coffee livelihoods can be extremely precarious and generate very poor social indicators, especially in those countries most dependent on coffee exports and most vulnerable to the chaos and uncertainty of the coffee rollercoaster. The small African country of Burundi, with a population over 8.5 million, is the most coffee dependent country in the world, relying on coffee exports for 54 percent of its total exports even while it only represents 0.2 percent of the world's coffee exports. In 2011, Burundi had a life expectancy at birth of only 50 years and a GDP per capita of $247, and was one of the world's lowest-ranked countries in the United Nations Human Development Index.[2]

Riding high above all of this are the transnational coffee roasters, retailers, traders, and supermarket chains, which rake in huge profits regardless of the ups and downs of the coffee rollercoaster. The global coffee chain is designed in such a way that, boom or bust, the lion's share of wealth flows comfortably into the hands of corporate giants like

Kraft Foods Group and Nestlé, the leaders in the global coffee industry and among the wealthiest companies in the world. During the darkest days of the global coffee crisis at the start of the new millennium, Oxfam International estimated that coffee farmers in Uganda received 14 US cents per kilo for green beans that were then transported, processed, and sold in supermarket chains in the UK as instant coffee for $26.40 per kilo – a price inflation of over 7,000 percent.[3] Needless to say, with a global value chain arranged as unevenly as this, the world's biggest coffee companies rode through the crisis without a scratch, continuing to garner hundreds of millions of dollars in profits while tens of thousands of coffee families confronted personal and community disaster.

In seeking to account for the causes and consequences of such an uneven global coffee economy, most analyses coming from official international institutions, nongovernmental organizations, think tanks, journalists, and policy advisors have tended to focus on the dynamics of the coffee market, its volatility and unpredictability and the ever-expanding oligopolistic dominance of roasters, retailers, and traders. This tendency has been further entrenched in recent years by the growing pervasiveness of fair trade, ethical trade, organic and sustainable coffee, corporate social responsibility, and any number of market-driven projects devoted to combating poverty and inequality. The result has been a dominant consensus around the "market" as the cause of underdevelopment and its potential solution, with the "state" receding ever further into the background. As an alternative, however, I will argue, building on some of the most insightful work on international political economy and the coffee industry, that the state and the market are inseparable and that *coffee statecraft*, both good and bad, has been and continues to be central to the everyday operations of the industry. While the global market does indeed cause socially destructive volatility and unpredictability for millions of coffee farmers, the existence of this market is not the natural or inevitable outcome of human activity, but rather stems from a specific form of international exchange set in motion by states that protect, reproduce, and contest it on a continual basis. The geopolitics of coffee statecraft must be of central concern for understanding and challenging the deep roots of uneven development in the coffee world.

Coffee statecraft

Coffee is the classic global commodity, produced as a generic or universal good specifically for sale on the market. It links the daily routine of millions of consumers and producers living thousands of kilometers apart and experiencing vastly different lives. The commodification of coffee, its processing, transporting, roasting, packaging, branding, and selling, generates immense wealth each year – indeed, in a world of high finance and financial crises, unprecedented technological change and cyber billionaires, it is easy to forget that a key source of global wealth is, literally, right under our noses. And yet, despite billions of dollars in profits made each year, the majority of the world's coffee families live in relative poverty. Who is the culprit for such vast inequality?

The most frequently named offender is the global coffee market. Nongovernmental organizations and social justice activists commonly point to the volatility and unpredictability of global coffee markets. These groups have, over the years, played a key role in exposing and condemning the lack of state "intervention" in coffee markets to assist the poorest farmers and workers. In recent years, however, concern for state intervention has been pushed aside and overshadowed by efforts to persuade corporations to voluntarily offer higher prices and support better labor and environmental standards in exchange for winning the loyalties of ethical consumers.[4] Leading this charge have been fair trade advocates, who have proclaimed that fair trade allows consumers to "vote with their dollars" and empowers them as "citizens" to "build mature markets and develop new ones, always continuing to innovate." They have moved increasingly away from state-centered visions of social justice and toward the belief that "Fair trade actually embraces many of the ideals of capitalist free trade; the distinction is that we aim to give everyone a fair chance."[5]

Trade economists, representatives of big coffee, and the most powerful states and international organizations, for their part, have generally condemned the idea of state intervention in the coffee market. In its place, they have proposed "free" market solutions that have long been on offer (increasing productivity and quality, expanding markets, diversifying . . .) and have continually failed to bring substantive benefits to the majority of small coffee farmers and workers.[6] Since the 1990s, they too have begun to embrace an array of sustainable coffee certification projects, from fair trade to organic to Rainforest Alliance. The World Bank, one of the most powerful international organizations in the world, now advocates that poorer countries pursue sustainable coffee certification, as long as it is combined with privatization, liberalization, and deregulation of the coffee sector.[7] Representatives of the mainstream coffee industry, for their part, have also gradually given way to modest support for sustainable certification, as long as coffee continues to be "bought and sold based on the free trade model."[8]

The growing consensus among otherwise opposing groups around the power of the "free" market to solve inequality and poverty through a burgeoning array of sustainable coffee projects has served to naturalize market domination. Both critics and defenders of the status quo have tended increasingly to focus their energies on the coffee market itself, as opposed to the states that create, regulate, and reproduce this market.

This direction of popular and political debate runs contrary to a great deal of scholarly debate on the topic, which often centers on the relationship between the state, the coffee industry, and the market at the local and global level. Existing scholarly works provide powerful insights into the nature of the coffee world, although most have little to say specifically about the geopolitics of coffee statecraft – how competitive capitalist states do battle for a bigger share of the global coffee pot, the outcome of which is immensely important to the livelihoods of millions of poor coffee farmers and rural workers globally.[9]

Two particularly important exceptions are the works of Robert Bates and John Talbot. Robert Bates' work on the political economy of coffee is of major importance because it draws attention to the ways in which states have shaped the international coffee market through various forms of market regulation (such as the International Coffee Agreement [ICA], discussed in chapter 3) in response to the demands of domestic coffee sectors. Bates tends to depict the state in fairly neutral terms, seeing it as a relatively benign institution swung in different directions by competing interest groups that vie for political power and influence. This book, in contrast, places greater emphasis on the *specifically capitalist and class nature of coffee states*. A key distinction is that while Bates depicts states as ultimately choosing to "introduce institutions into economic life," I argue that the state is always involved in managing coffee markets, under both "regulated" and ostensibly "free trade" regimes.[10]

John Talbot's path-breaking work on coffee and the ICA, drawing on a global value chain approach, more effectively gets at the capitalist drives underpinning the global coffee industry. Whereas conventional economic trade theory tends to assume that trade participants are independent from one another, connected by only isolated economic transactions, the global value chain approach points out that officially independent firms are linked through informal institutional frameworks coordinated by "lead firms" that "govern" the commodity chain based on their economic dominance over such things as market access and information. Using this approach, Talbot has effectively demonstrated how "collective action" among producing states was able to provide higher and more stable prices than was the case before or after the ICA.[11]

The argument I provide here around coffee statecraft draws on Talbot's insights, while making divergences from the global value chain approach, which has been criticized for an overemphasis on *economic* agents, in particular transnational corporations (TNCs), at the cost of downplaying the significance of other actors, such as states, social movements, and international regimes.[12] The state, in particular, is not just one player among others fighting for its share of the wealth, but is central to creating and reproducing the very social relations that underpin the entire global coffee chain. Much global value chain work tends to make too firm a distinction between when a chain is dominated by economic and when by political governance, whereas both are always present.

Building on and departing from these works, I argue that the complex relationship between the modern state and the global coffee market can be further understood through the idea of "coffee statecraft" – a term drawn from Peter Gowan's insightful use of "economic statecraft." Of particular importance is David Harvey's understanding of the geostrategic interests of capitalist states being primarily driven by two logics, a "territorial logic" and a "capital logic," intertwined in a relationship that can be compatible, competing, or contradictory, depending on the specific context. The "territorial logic" involves the considerations of political leaders as they seek power for their state over other states. For coffee exporting countries, this includes defending and promoting the interests of their national coffee sector, essential for the state in providing revenue (through tariffs and other forms of taxation), employment, economic and social stability, and institutional legitimacy. The coffee sector, in turn, exercises influence over state priorities depending on their overall economic weight, control of major media and information outlets, and direct influence over government through political contributions, donations, and bribes. Importantly, territorial logic is not confined to direct political or military activities, but also entails Gowan's notion of "economic statecraft": the strategic use of market management by capitalist states to gain advantage or power over others.[13]

The "capital logic," in contrast, involves state action to develop, protect, and reproduce on a global scale a capitalist economy, based on specific social relations and commodity production. This includes protecting private property, managing class conflict (through coercion or social reform), defending the interests of capital at home and abroad, and, as Ellen Meiksins Wood has argued, ensuring the artificial separation between the economic and political realms required for the capitalist economy to function.[14] State and state policy makers in the modern world are far from autonomous, but rather must constantly respond to the imperatives of the economic system upon which the state is built. This means that the capitalist state must work to ensure the relatively smooth functioning of the capitalist economy – which is never guaranteed. It also means that the state must disproportionately respond to the needs of dominant classes – local and transnational capital that possess immense economic power well beyond that of mere "special interest groups."

The relationship between the "territorial" and "capital" logics can be relatively harmonious or conflict-ridden, depending on the particular context. As we will see in the chapters that follow, the history of coffee is replete with examples of coffee states employing authoritarian violence to defend highly unequal distributions of land and resources in the interests of domestic agrarian elites and transnational companies (the capital logic), only to be confronted by social protest, social revolution, and civil war from below (significantly disrupting the territorial logic). In other cases, states have been able to more effectively manage the competing logics, as has been the case in Costa Rica, where short-term pain imposed on the agro-industrial elite (through taxation and other forms of state management) ultimately resulted in a highly efficient coffee economy and significant social welfare gains for broad sectors in society.

The coffee market

On this framework, this book sets out to demonstrate how coffee statecraft plays a key role in driving the global coffee industry. States not only intervene in markets more than we might think but set the very context in which markets operate: enforcing private property; managing (or mismanaging) social conflict and political stability; providing countless rules and regulations concerning everything from legal contracts, domestic currency, foreign exchange, and international treaties to labor rights and family law; and constructing a wide range of public infrastructure, from roads and telephone lines to schools and water systems. States and markets in a capitalist economy are inextricably intertwined. This book seeks to reassert the centrality of states to the global coffee economy, while not losing sight of the fact that global markets underpin major economic imperatives and private institutions (such as the TNC) that can be more powerful than individual states, depending on the state in question – Burundi is obviously much less able to impact global markets and giant transnational companies than is the United States or Brazil. Two characteristics of the global market, in particular, have been particularly dominant in the coffee world and are widely discussed and debated: extreme market volatility, and intense corporate oligopoly.

Market volatility and the coffee rollercoaster

Coffee's trade and production pattern strongly exemplifies one of the key problems facing most tropical commodity industries: price volatility and declining terms of trade compared to manufactured goods. This leads to intense cycles of boom and bust, causing periodic crises for small farmers and rural workers. Conventional trade economists have tended to view these cycles as distortions in the market, which is assumed to move toward equilibrium if left unhindered. They are, however, systemic to most commodity trading due to the need for individual farmers to produce goods for the market without certainty that buyers will exist to purchase them, or that competitors will not absorb the demand that they thought existed earlier.[15]

The global coffee market has been particularly subject to the "commodity problem" due to the specific nature of coffee cultivation, especially for the most popular Arabica beans. Unlike many other crops, such as wheat and corn, Arabica beans are grown on perennial plants that take three to five years to mature and require a major commitment of capital. During the boom, prices are artificially high as new trees mature, leading farmers to plant even more trees in expectation of continued high prices. This sets the stage for an eventual glut in supply and a bust in prices, by which time farmers are unable to easily switch to other commodities due to relatively large amounts of fixed capital they have invested in the coffee trees.[16] This extreme boom and bust cycle has made the global coffee chain the frequent target of an array of state and non-state-led development initiatives (perhaps more than any other global commodity) as well as the site of intense levels of corporate concentration and control.

Corporate oligopoly and coffee giants

A process of intense corporate oligopoly has mirrored market volatility in the coffee chain. Unpredictable market patterns make it very difficult for smaller companies to survive while giving incentives to larger companies to grow ever bigger, enhancing their ability to profit despite market downturns. Northern-based transnational coffee retailers and roasters at the top of the global coffee chain have the ability to manipulate global prices in their favor, artificially increasing the gap between the relatively low "farm gate" price paid to farmers for unprocessed green beans and the higher prices paid for roasted beans on retail markets. Coffee giants possess significant advantages over smaller competitors due to their economies of scale, massive marketing budgets and "brand power," access to new technologies and innovations, and political influence, required for promoting domestic and international policies favorable to corporate interests.

Far from a world of free trade, the unstable and competitive nature of the global coffee market has resulted in a long historical process of bankruptcies, mergers, and acquisitions that has eroded competition and shifted volatility into the hands of the poorest and most vulnerable workers and farmers. During hundreds of years of colonialism and slavery beginning in the fifteenth century (discussed in chapter 2), the global coffee trade was created and expanded by large colonial companies granted official trade monopolies by their imperialist home states in Europe. Starting in the nineteenth century, these companies were broken up in favor of private corporations that gained ever more control over the import, processing, and distribution of coffee. Throughout the twentieth century, corporations used their economic power to expand coffee consumption by ensuring a steady supply of cheap coffee beans and conducting multimillion-dollar marketing campaigns designed to persuade people to drink more and more coffee. The high costs of these efforts spurred further corporate concentration: in 1915, there were over 3,500 roasters in the United States; by 1965, there were only 240, and the top eight accounted for 75 percent of all coffee sales.[17]

Today, the global coffee chain is dominated by immense trading companies that frequently trade in multiple commodities; oligopolistic supermarket chains that dominate access to major consumer markets; and, above all, transnational roasters that coordinate the processing and distribution of the finished product. Roasters have been consolidating their control over the global industry, especially since the 1990s, developing special relationships with major retail supermarket chains, taking ownership of their own trading companies, and developing direct arrangements with enormous Southern coffee plantations based on private contracts. Five roasters now dominate the global coffee chain, accounting for nearly half of the world's supply of green beans: Kraft Foods Group buys just over 13 percent of the world's total, followed by Nestlé (13 percent), Sara Lee (10 percent), Procter & Gamble (4 percent), and Tchibo (4 percent). The largest companies control multiple coffee brands, manage countless products beyond coffee, and possess immense economic weight. Nestlé, for example, is the world's largest food company, with combined assets in 2012 worth over $122 billion, roughly equivalent to the GDP of the world's second largest coffee exporter, Vietnam, and its 88 million inhabitants.[18]

Trends toward corporate oligarchy have been partially diverted since the early 1980s with a boom in specialty coffee sales, which now account for over one third of the US coffee market, the largest single coffee market in the world. This has spurred the growth of small, independent coffee houses, so that there are currently around 1,200 roasters in the United States. The historical pattern toward corporate concentration, however, is increasingly apparent within the specialty coffee industry. Starbucks Coffee Company, the largest specialty roaster in the world, began with a single store in 1971 and now operates over 19,000 stores in 62 countries. Starbucks' net revenue in 2013 for its "Americas" segment (including stores in the United States, Canada, and Latin America) was equivalent to around one-third of the retail value of the entire US coffee market.[19]

The flip side of corporate concentration and power at the top of the coffee chain is the vulnerability and lack of power experienced by the majority of the world's coffee growing families in the South. To highly varying degrees depending on the country in question, coffee growing is dominated by powerful classes of plantation owners and agro-processing elites that possess extensive transportation and processing infrastructure; employ low-paid, highly vulnerable workers; have considerable political influence over local and national states; and utilize their economies of scale to make profits even under price conditions that would otherwise be highly destructive. The vast majority of the world's coffee, however, comes from small farmers who lack market power, sufficient resources, or significant political influence, and seldom have viable "exit strategies" into other livelihood options. They are highly vulnerable to the demands of the global market and the coffee giants that dominate processing, shipping, and distribution at the international, national, and local level. This extremely uneven process – a world apart from a neutral exchange between willing and autonomous market agents – ensures great wealth for a few and relative poverty for the majority.

Coffee and "free trade"

In seeking to account for uneven development and inequality in the coffee world, much of the popular debate has been
framed around issues of "free trade," with opponents and
proponents arguing for or against it. In this book, however,
I will argue that there has never been any genuine "free
trade" within the coffee industry to speak either in favor of
or against – it is, in many ways, a "red herring" in seeking to
understand and address poverty and inequality in the coffee
world.

The idea of free trade has a long history dating back hundreds of years. Its recent popularity, however, has been driven
by the rise of neoliberal policies to hegemonic status among
wealthy states and international financial institutions beginning in the 1970s. Neoliberals are opposed to what they see as
state "intervention" in the economy, which they feel distorts
supply and demand, wrecks the market signals they produce,
and leads to inefficiency and waste. Consequently, they call
for the state to remove itself from the market and let it operate "freely." The notion that this can even take place, however,
requires a leap of faith far removed from the actual history
of capitalism and world trade. Throughout modern history,
the capitalist state has played a central role in creating the
conditions for a capitalist economy to operate. Moreover, the
state has been particularly present among the most successful capitalist economies. It has been widely documented that
both rich Northern nations and rapidly growing Southern
economies today have all developed with a variety of protec-
tive mechanisms in place, including import controls, tariffs,
levies, quotas, and preferences designed to protect domestic
and promote export industries.

Given the messy history of capitalism and the role of the
state, the ideas around removing the state from the market

are generally premised on economic formulas and mathematical modeling based on what a free trade world *should* or *could* look like. Real world trade is then measured up against these models to determine how various "imperfections" can be eliminated. This approach, grounded in "neoclassical economics," has long been critiqued for being based on an array of speculative foundations, including the assumption that people are primarily self-centered, individualist, and competitive, and that the market is always moving toward an eventual state of "equilibrium" – as opposed to constant change and frequent chaos.[20] Regarding trade specifically, such economic modeling has been used to assert and defend the benefits of "free trade" and the search for "comparative advantage," while neglecting a variety of real-world factors, including the many ways in which the ideal of comparative advantage is hindered or blocked entirely by imperfect information, the limits of existing technology, geography, historical path dependence, the fixed cost of existing infrastructure, and the skill set of the existing workforce.

Moreover, while free traders want to see politics removed from trade, very often politics has played the key role in formulating and reproducing actual trade patterns. Historically, colonial powers imposed trade relations on subjugated regions through direct political violence to the benefit of imperial states, many of which proclaimed themselves to be "free traders," and the disadvantage of colonized states in the South. In the contemporary context, politics continues to be a driving force behind trade and trade agreements. This includes the preponderance of "free trade" agreements, which have been widely criticized for being ineffective at evenly eliminating trade barriers while containing components that go well beyond trade, such as extensive protections for intellectual property and corporate investment rights. Well-known economists from different political positions have noticed this discrepancy, calling for genuine multilateral free trade and blaming "ideology" and "politics" for distorting economic policies that are sound *in theory*.[21]

And yet, while noting the distorting effects of politics and ideology, economists in general have continued to pursue and promote economic trade theory with these factors left out, or introduced on the sidelines as "imperfections." Politics and ideology, however, are central to human life and cannot be separated from economics in this way. "Free trade" is in fact central to understanding the global coffee industry, but not from the vantage point of mainstream economics. Rather than being merely an objective technical or policy issue, "free trade" is a complicated political, economic, and ideological "package" rooted in complex social, historical, and cultural forces.[22] Politics and ideology are not side issues for free trade, but rather central to it, often more important than the genuine free trade of goods, which generally does not exist in world trade. Addressing this, Jodi Dean has developed the idea of the "free trade fantasy," reproduced through the efforts of the corporate media and of rich and powerful research centers and "think tanks" that produce "knowledge" biased in favor of the interests of economic and political elites and dominant states. The power of the free trade fantasy, argues Dean, is to "link together a set of often conflicting and contradictory promises for enjoyment and explanations for its lack (for people's failure to enjoy despite all of the promises that they would)." Despite history and current events revealing major oversights in the expectations of free traders, proponents continue to insist that these instances are mere deviations from how the world *should* operate, fulfilling what Dean refers to as "the 'excuses, excuses' role of fantasy." The free trade fantasy fills in for the lack of actually existing "free trade."[23]

Nowhere is this fantasy more apparent than in debates about the global coffee industry, where it is frequently assumed that the current conditions of the industry are the outcome of market forces of supply and demand (as opposed to power and politics) and that the economics of coffee can be understood with politics and ideology left out.

The politics of coffee

To challenge the dominant understandings of the free trade package, this book offers a political assessment of what would often be seen as primarily economic forces at work in the coffee industry. The chapters of the book are arranged with this goal in mind. Chapter 2 offers a brief history of coffee from colonial times until today. Its focus is the long historical process through which coffee became a major global commodity, driven by conscious effort on the part of colonial and imperialist states. The process involved creating the original conditions for a highly unequal system of coffee production, trade, and consumption and ensuring that these conditions were maintained, often through brutal force. States have not only protected – and in some cases opposed – unequal distribution of coffee land and resources, but have also worked to ensure a specific set of social relations around commodity production for export and the separation of the economic and political spheres, a central characteristic of a capitalist economy. On the basis of this history, I argue that the economic and political inequality that plays itself out in the coffee industry today cannot be resolved through market adjustments alone, but rather requires attention to the ways in which coffee statecraft has forged these deep structural roots.

Chapter 3 focuses on the rise and fall of the ICA, from 1963 to 1989, a quota system signed by all major coffee producing and consuming countries, designed to stabilize and increase coffee prices by holding a certain amount of coffee beans off the global market to avoid oversupply. Given the "imperfect" market conditions of the real-world coffee economy, I argue that the ICA, despite its shortcomings, was superior to the official "free trade" regimes that preceded and followed it. The ICA resulted in higher incomes for coffee farmers and offered greater "social efficiency," measured not strictly on the basis of a narrow definition of economic efficiency, but on the broader needs of society as a whole.[24] Both "regulated" and "free trade" coffee markets have involved significant state management, only some management has been more "socially efficient" than others.

In chapter 4, I discuss the current era of "free trade" in the coffee industry and argue that coffee statecraft has continued to play a central role in shaping the global coffee market. The chapter focuses on the worst years of the global coffee crisis, from 1998 to 2002. The crisis is frequently understood as the outcome of the decline of state regulation and the unleashing of market forces after the collapse of the ICA. While I agree with the overall assessment of the impact of the end of the ICA, I argue that economic statecraft remained a primary driver of the coffee market before, during, and after the crisis. It is generally noted that one of the primary causes of the crisis was the rapid entrance of newcomers into the coffee market, especially Vietnam, which developed from an insignificant coffee exporter to the world's second largest by the end of the 1990s. This did not occur as a result of spontaneous market forces of supply and demand, however, but out of a conscious effort by the Vietnamese state to promote coffee production and export – engaging in coffee statecraft to meet its own territorial and capital logics, with mixed success. The end of the ICA did not mark the beginnings of a golden era of free trade, but rather a shift from a degree of collective action among coffee states to intensified competition between them. Increased competition remains to a significant degree driven and managed by coffee statecraft, in a manner that sparked the global coffee crisis and continues to drive coffee markets to this day.

Chapter 5 shifts the focus to two significant trends in major Northern consumer markets since the late 1970s: the growing power of giant roasters and retailers, which have pushed the industrialization, mass distribution, and homogenous consumption of international brands; and the growth of the specialty coffee industry, which has offered niche markets for those seeking better ethical, environmental, and health standards, including various forms of fair trade, organic, and sustainable coffee. Much has been written in recent years assessing the strengths and weaknesses of fair trade and corporate social responsibility. In this chapter, I summarize some of the major conclusions of this work, paying specific attention to the growing convergence and conflict between the two market-driven projects. While fair trade has historically been a "bottom-up" project whose participants have been highly critical of "top-down" corporate social responsibility, since the 2000s there has been a growing convergence between the two as corporations have begun to give token support to fair trade – in Starbuck's case, around 8.1 percent of their beans were certified fair trade in 2012 – as part of their overall corporate branding strategy.[25] This convergence has at times broken out into significant conflict; in 2011, Fair Trade USA broke with the rest of the fair trade system and now seeks to develop itself as a more pro-corporate global certification body. The chapter concludes reflecting on the limits of both projects, which naturalize market rule, offer a form of neoliberal privatized governance that cannot match previous efforts at social regulation by states, and obscure the continued pervasiveness of coffee statecraft in managing the coffee chain.

The final chapter highlights some of the new and emerging trends within the global coffee industry. Competing states continue to conduct coffee statecraft to meet the needs of their territorial and capitalist logics in varied ways, only in a manner that has turned away from more robust visions of the "developmental state" in favor of a "non-developmental state" that perpetually denies the ability of the state to act in the broader social interest. Consequently, the dominant historical patterns within the industry have remained more or less intact, even while significantly impacted by emerging trends in the world system, including the cost-price squeeze, the financialization of coffee, and the intensification of environmental crises. Perhaps the newest and most unanticipated trend has been the "rise of the South," with a number of larger Southern countries experiencing unprecedented rates of economic growth and gradual improvement in most major social indicators. This has sparked new forms of cooperation and competition within the coffee industry, and afforded poorer and more vulnerable countries greater policy and ideological space to rethink coffee statecraft in subtle but important ways, as evidenced by the Ethiopian trademarking initiative and by the rise of fair trade South, headed by the Latin American and Caribbean Network of Small Fair Trade Producers (Coordinadora Latinoamericana y del Caribe de Pequeños Productores de Comercio Justo, or CLAC).

At the end, I reflect on the alarming coffee leaf rust crisis in Central America, which has cost hundreds of thousands of jobs and hundreds of millions of dollars in lost income. The roots of the crisis are not about coffee leaf rust alone, but are deeply interwoven with coffee's everyday crises of inequality, poverty, and vulnerability. Ultimately, major changes to the coffee industry are required that would put more income and resources into the hands of farmers and workers through state-supported education, health, and agricultural programs, land reform, and a revival of international price regulation.

Conclusion

Extreme market volatility and corporate oligarchy remain two central pillars of the global coffee industry. These pillars, however, have underneath them a foundation: a highly uneven world system, cut through by global and domestic economic, social, and political inequalities, ultimately underpinned by the power of states in an interstate system. In this book, I emphasize the geopolitics of coffee statecraft to demonstrate the central role states have played in the shaping of the global coffee economy, even in the era of "free trade"; the pitfalls of neglecting, overlooking, or downplaying this role; and the possibilities for compelling states to adopt more socially and ecologically just forms of coffee statecraft – possibilities that *do* exist, however difficult it might seem.

CHAPTER TWO

Making coffee

Despite coffee's special place in the hearts of consumers as a source of warmth and comfort, it has often been observed that coffee has a long, dark history, and is grown under conditions far removed from the upbeat and sexy imagery employed on café walls and in multimillion-dollar advertising campaigns. Making coffee requires far more than scooping beans into a coffee machine, or the more elaborate ceremonies of coffee aficionados. Long before roasters turn coffee into the dark brown beans familiar to most consumers, green coffee beans must be grown, processed, traded, and shipped; but where, by whom, and under what conditions? The long historical process through which these decisions were made was a world apart from the free trade ideal of a neutral market exchange between relatively equal market agents, and the political process that guided it was not generally democratic, open, or participatory.

The historical process through which coffee became a major global commodity bears heavily on the world of coffee today. Coffee statecraft played a significant role in this process, especially statecraft conducted by colonial and imperialist powers that developed and policed an unequal system of coffee production and exchange, often through brutal violence aimed at the local population or at weaker, subordinated states. States have not only protected inequalities in power and resources in the coffee industry, but have also worked to ensure a specific set of social relations around commodity production for export and the artificial separation of the economic from the political that is central to a capitalist economy. This historical argument differs from the free trade package, which tends to downplay history and its relevance to the current state of the world. Two trends particularly dominant in conventional economic thinking have immense significance for how trade policy is envisioned in the coffee world and beyond.

First, in place of systematic history, trade economists often offer vague depictions of capitalism as the inevitable outcome of the natural human desire to trade freely in a market economy. The 99 percent of human history under which the world lived prior to capitalism (the vast majority as hunting and gathering communities) is seen as the lead up toward the realization of people's true nature today. This vision of capitalist history tends to overlook and distort some of its central characteristics. Capitalism is depicted as being primarily about trade, as opposed to a specific set of social relations that underpin production and trade. The capitalist market, as Ellen Meiksins Wood has observed, is depicted entirely as a place of "opportunities," as opposed to also being a place of "imperatives" – to find a job, to stave off bankruptcy, to produce more at an ever-faster pace, to exploit finite resources. And finally, the modern state is portrayed as the antithesis to the market, whereas in fact capitalism has given birth to the largest states in human history to codify and regulate the "unregulated" market.[1]

Second, mainstream economists have tended to not only offer a thin understanding of history, but also repudiate its relevance. Renowned economist Jeffrey Sachs, in his highly influential book *The End of Poverty*, morally condemns the brutality and racism of the colonial era, but suggests there is no basis for the assertion that "the rich have gotten rich *because* the poor have gotten poor." Over the past 200 years, both rich and poor regions have experienced increased economic growth, and in today's world the "[b]ad policies of the past can be corrected. The colonial era is truly finished."[2] The effect is to turn the past into a strictly moral affair – one at which we can all *now* wag our fingers – while denying its continuing relevance to this day.

Yet, while all countries may have seen some gains in economic growth over the past 200 years, this tells us little about how different countries have been inserted into the global economy in very different ways – with dominant states managing dynamic, highly industrialized and high-technology economies, while former colonies have been left with weak, vulnerable economies frequently dependent on the export of commodities of lesser economic value. Those who have received only a shred of the economic gains of the past 200 years are expected to now compete against the economic giants that have taken the lion's share of the wealth. Thus Uganda, which relies on coffee exports for 20–30 percent of all foreign exchange earnings, and has a GDP per capita of $470 and a life expectancy at birth of 54 years, is expected to compete in the global economy with its former colonial master, the United Kingdom, which has a GDP per capita of $38,818 (over 82 times that of Uganda) and a life expectancy of 80 years (26 years more than Uganda, significant not only in moral terms, but economically and politically as well). Far from a lingering shadow of the past, the history of coffee must be central to our understandings of the coffee world today.

A brief history of colonial coffee

Coffee became a major global commodity through a long and complicated historical process lasting centuries and entailing the gradual expansion of production and consumption, alongside new developments in international transportation, shipping, processing, and grading. Whereas North America and Europe emerged as the core coffee consuming regions in the nineteenth century, for nearly three hundred years prior to that, the expansion of the international coffee trade was dominated by Indian and Arab merchants linking growers in Ethiopia and Yemen with major markets in the Middle East and North Africa. Yemen in particular dominated the production of coffee beans for export from the fifteenth to the eighteenth century. Coffee was grown in Yemen predominantly by peasants who combined farming subsistence crops with growing small amounts of coffee beans for sale on the market. The main coffee consumers were middle-class Muslims in regions where alcohol consumption was prohibited as a result of Islamic laws, making coffee all the more desirable. Through the gradual expansion of production and consumption centered largely on North Africa and the Middle East, coffee first became a global commodity integrated into the expanding world trading system.[3]

It was not until the middle of the seventeenth century that coffee gained notable popularity in Europe, although mostly as a luxury crop for wealthy consumers. Muslim countries continued to dominate coffee consumption well into the eighteenth century. Europe did, however, begin to play a more central role in meeting the demand for coffee. The Dutch Empire was the first European power to get significantly involved in coffee, initiating coffee growing in the colony of Ceylon (today Sri Lanka) in the 1650s and then expanding to Java, Sumatra, and other colonies in the South and South East Asia by the end of the century. Coffee was grown primarily by local peasants, each of whom was forced by decree to plant several hundred coffee trees on their land and sell the beans to the Dutch East India Company at set (and very low) prices. The combination of expanded cultivation and exploited peasant labor allowed the Dutch to expand their penetration of Middle Eastern and North African coffee markets while also encouraging the growth of coffee markets in Europe aimed at the emerging urban middle classes.

Demand for coffee continued to grow throughout the eighteenth century, part and parcel of the general expansion in demand for a range of tropical commodities. European colonial states played an ever-increasing role in this expansion, granting imperial monopolies along with economic and military support to mercantilist trading companies that extended colonial domination and expanded the Atlantic slave trade.[4] Originally, the slave trade had been developed by the Portuguese Empire in the fifteenth and sixteenth centuries to provide hyper-exploited labor for colonial sugar plantations. The slave trade grew substantially in the eighteenth century in response to the increasing demand for tropical commodities. In the eighteenth century alone, European colonial powers forcibly exported around six million African slaves, more than three and a half times the number of the previous two and a half centuries.[5] The British Empire took the lead in the slave trade at this time, emerging as the world's largest slave trader and profiting substantially from the cheap commodities that helped fuel industrialization back home.

While the British Empire benefited significantly from the slave trade, as well as from its general colonial conquests, these imperial activities did not in and of themselves spark the industrialization in England that would come to transform the world system. Imperial Spain, for example, in the eighteenth century had been a dominant colonial power for more than 300 years, yet did not emerge as the leader in industrial capitalism. For the large-scale and rapid industrialization such as that experienced in England to occur, a fundamental transformation had to take place in the mode of production, involving a shift from a feudal to a capitalist one. Feudalism, which had dominated Europe for centuries, centered on a system wherein the elite extracted surplus wealth from the peasantry, or from conquered territories, through direct military coercion. This required endless expenditure on state building and warfare, reducing the resources available for investing in consumption and production and resulting in limited productivity.

With capitalism, on the contrary, competitive pressures drove historically unprecedented rates of economic growth and industrialization by pushing all market agents toward increased productivity, technological innovation, and the relatively efficient investment of surplus wealth. The emergence of a new system of land tenure in England in the sixteenth century laid the basis for the unique conditions that would, over centuries, bring the transformation from feudalism to capitalism about. Whereas feudal peasants had formerly been tied to the land, devoting their energies primarily to subsistence farming, under the new rules they could now be dispossessed if they failed to make rent payments in currency. This meant that peasants had to produce for sale on the market, sparking new competitive pressures among them. As land became concentrated into fewer and fewer hands – those of the most successful farmers – thousands of former peasants flocked to the urban cities to sell their labor to survive. In doing so, they became the new industrial working class that was central to the capitalist transition underway in England by the end of the eighteenth century.[6]

Coffee and capitalism

More than just laborers, the urban working class, no longer producing food and materials at home for their own needs, formed a new mass consumer market in everyday goods. Coffee became a particularly popular product for this new market, well suited to the demands of the new factory system. With men, women, and children compelled to work at the factory for long hours with little time at home, coffee's appeal stemmed from the fact that it could be prepared quickly and would not easily spoil. Coffee was also of great appeal to the capitalist classes, concerned as they were with a disciplined and controllable workforce, and so they welcomed coffee as a replacement to the more socially disruptive alcohol.

The growing popularity of coffee in Europe intensified slavery and imperialist statecraft abroad, as well as resistance to it. While Yemen remained the world's largest coffee exporter until the middle of the eighteenth century, by the 1770s it had been eclipsed by the French Empire, which provided half the world's coffee exports primarily through the forced labor of nearly half a million slaves on the colony of Haiti (then Saint Domingue). This dominance was brought to an end through a massive slave revolt beginning in 1791 and ending in 1804, with former slaves victoriously declaring an independent Haiti under black rule. The costs of the victory, however, were great, and continued aggression by colonial powers, combined with the huge economic, political, and social toll of the liberation struggle, wrecked the island's export-dependent economy. This set the stage for long-term decline, the tragic impact of which is still felt today in Haiti, which continues to be one of the poorest countries in the world.

While Haitian independence put an end to the French Empire's coffee dominance, the international coffee economy was just hitting its stride. The nineteenth century witnessed a major boom in coffee production and export throughout the South. Brazil, which gained independence from the Portuguese Empire in 1822, quickly emerged as the new undisputed leader of the global coffee industry. This marked a shift in Brazil away from sugar, which had been a dominant crop during the colonial period, grown on enormous *fazendas* (plantations) owned by a tiny agrarian elite. The elite ruled politically and economically over masses of African slaves who labored under terrible conditions, living an average of seven years after being "imported," as it was considered cheaper to buy new slaves than to provide for the needs of existing ones. These conditions remained largely the same after independence, with Brazilian slave owners changing the crop, transitioning toward coffee in response to declining sugar prices.

In fact, as coffee growing in independent Brazil increased, so did the import of African slaves, with the slave population reaching over two million by the middle of the nineteenth century. Brazil was eventually compelled to end slave trading under pressure from the British, who formally abolished the slave trade in 1807. Domestic slavery, however, continued in Brazil until 1888, by which time Brazilian plantations had come to account for nearly half the world's coffee production. Following Brazil's lead, other countries picked up production to take advantage of coffee's growing popularity, including Java, Ceylon, and the countries of Central America. Combined, these countries drove a glut in the global coffee market, with supply outpacing demand, driving prices down, which in turn increased its accessibility and popularity. In the United States, lower prices were facilitated by the fact that it was the only major coffee consuming country to import coffee tax-free for most years from 1832. By the end of the nineteenth century, the United States accounted for 40 percent of world coffee imports. Consumption also increased substantially in Europe, especially in Northern countries. The rapid increase in Brazilian production combined with equally impressive increases in North American and European consumption fundamentally altered the world coffee trade during a period referred to by historian Steven Topik as "coffee's heroic nineteenth century."[7]

Brazil not only led the world in coffee production in the nineteenth century, but also pioneered many of the planting, cultivation, and processing methods that still dominate much of the coffee world to this day. At the time, Brazil grew strictly *Arabica* beans, which were the only beans grown for export until the twentieth century, and remain the most widely exported. In terms of cultivation, Brazil was the first to carry out "full-sun" coffee growing on a massive scale. This involves clear-cutting all existing forestry to make way for coffee trees to be lined side by side in rows up and down hills. The sun-exposed trees grow rapidly, which is the method's great advantage, but only at the expense of intensified soil erosion, the depletion of soil nutrients, and the elimination of natural predators to coffee pests and disease. Coffee trees are planted as monocultures, with a limited number of similar crop variants, resulting in Arabica trees coming from a narrow genetic stock and being highly susceptible to diseases, fungi, bacteria, and pests – most notably, coffee leaf rust (a fungus) and the coffee berry borer worm or *la broca* (a pest). Consequently, the full-sun method has generally resulted in escalating reliance on chemical fertilizers, pesticides, and insecticides.

Not only has Brazil had to deal with the ecological impacts of full-sun, monoculture cultivation, but much of the country has also not been climatically well suited for coffee growing. The ideal coffee lands for Arabica beans are generally considered to be between 3,000 and 6,500 feet above sea level in warm areas with an average temperature between 17 and 25 degrees Celsius. Much of Brazil suffers from periodic frosts, which have ruined almost entire coffee harvests on many occasions, and the majority of the country is below 3,000 feet, resulting in beans that are considered to be of lower quality. Despite these shortcomings, Brazilian coffee has trudged on, driven by full-sun, high-input farming that has set in motion a long historical pattern where planters have exhausted existing lands and then moved on to clear-cut and plant new farms, often in the country's ecologically sensitive and biologically diverse tropical rainforests. By the middle of the nineteenth century, for example, coffee farmers had exhausted the land in the Paraiba valley in Brazil, so they packed up and moved to the southern and western regions of the state of São Paulo.

As coffee production and export expanded, coffee came to play a central role in Brazil's further integration into the global capitalist economy. This did not at first entail the development of a large working class and wage labor, as was the case in industrial England. Brazil's coffee success began on the back of slavery. As slavery subsided, the agrarian elite turned toward debt peonage, wherein European immigrants were loaned the costs of passage to Brazil and a piece of land, but had to work under conditions of near-slavery until all debts were paid – from 1884 to 1914, over one million Italian immigrants came to Brazil in this way. The key component of Brazil's integration was not free labor, but intensified dependence on the production of commodities for sale on the global market. Previous forms of agricultural production, whether large-scale or small-, had involved maintaining significant land and resources for growing food for local consumption. The expansion of coffee brought with it a significant transition toward monoculture plantations devoted almost entirely to gaining income through the export of commodities to distant markets.

This general trend was experienced throughout much of the coffee world, although in diverse ways, where pre-capitalist forms of production were pervasive in laying the groundwork for a capitalist economy in the nineteenth and early twentieth centuries. Throughout much of Central America, for example, many indigenous groups were self-sufficient and entirely uninterested in abandoning their plots for the vulnerability of wage labor. As a result, emerging capitalist states often forced them to do so, stealing communal indigenous lands and compelling them into slavery, debt peonage, or highly onerous tenant relations. A characteristic case occurred in Guatemala from 1873 to 1885 under the Liberal dictatorship of Justo Rufino Barrios. Barrios declared as "idle" all of the land not devoted to pasture or export crops, predominantly indigenous land, seized it, and handed it over to the powerful agrarian elite. To force Maya indigenous groups to work on the new plantations, Barrios employed direct coercion, reviving the colonial-era forced labor system.

Through this process, powerful agrarian elites possessing enormous coffee plantations rose to economic and political dominance throughout much of the coffee world. Smallholder cultivation, however, persisted as the predominant form of coffee landholding (the majority of the world's coffee farmers are small-scale); as a key source of coffee supply for the global industry; and, frequently, as a major source of compulsory or semi-subsistence labor for nearby giant plantations.[8] Under certain conditions, smallholders even emerged as the dominant coffee producers in the country. In Costa Rica and Colombia, for example, the relative lack of large stretches of available land, capital, and cheap labor resulted in coffee production being dominated by small and medium-sized farms, tenant farmers, or sharecroppers. These farmers, drawing on family labor and providing many of their own subsistence needs, could survive and sometimes thrive under conditions deemed unprofitable by agrarian elites.[9]

Whereas giant Brazilian planters pioneered full-sun coffee growing, small and medium farmers in Central America and Colombia played a central role in developing and extending the "shade-grown" method – although either cultivation method can be employed on farms of varying scale. Shade-grown farming involves growing coffee under the shade of the forest canopy, trimming the branches to allow in sunlight. While trees do not produce as many beans as under full-sun conditions, the soil is protected from erosion and nutrition loss. Shade-grown also avoids the high costs associated with chemical-intensive farming, and as a result has been and continues to be the method of choice for the majority of small farmers in Latin America. Since the early 1970s, however, many of the more developed coffee economies, including Costa Rica and Colombia, have turned to full-sun growing on a massive scale.

It is not just the cultivation method that varies significantly between and within coffee countries, but the primary processing method as well, which is considered to have the greatest impact on bean quality. In Brazil from the nineteenth century until today, the dominant processing method has been the "dry method." This involves stripping coffee branches of all their cherries, which are then spread out on patios, turned several times a day, and allowed to dry and harden. The dry husks are then pounded off, leaving a green bean, which is then sized and polished before being exported. The method is fast and efficient, but generally considered to produce poorer quality beans, as both ripe and unripe beans are harvested together and beans can absorb flavors from the ground or get moldy lying on the patios. Today, around 90 percent of Brazil's Arabica beans continue to be processed through the dry method, which is also common in other parts of the coffee world, including Ethiopia, Haiti, and Paraguay.

The other widely adopted primary processing technique is the "wet method," which involves handpicking coffee cherries and immersing them in water, allowing the unripe beans to float to the top and be removed. The skin and pulp of the cherries are then removed in manual or gas-powered depulping machines, followed by a 24- to 48-hour fermentation process in large water tanks. If done carelessly, this stage can have negative ecological impacts, if the removed pulp is allowed to float carelessly downstream or if the wastewater is expelled back into nearby streams, which can lead to reduced oxygen levels in water and threaten aquatic life.[10] Once depulped, the remaining green beans are dried in the sun or in large, heated cylinders, and broken, black, moldy, or over-fermented beans are removed. The wet method can be extremely labor intensive, but typically results in higher-quality beans. Consequently, it has been widely adopted by small farmers seeking a competitive advantage against large plantations with economies of scale. The wet method was first taken up on a large scale in Central America and Colombia and has since spread to Mexico, Peru, Kenya, and the rest of the coffee world. Regardless of whether the wet or dry method is chosen, the green beans have traditionally been the main final product exported from the South, with large-scale roasters in the North dominating the roasting, retailing, and high-tech processing of green beans into the brown whole, ground, or instant coffee recognizable to most consumers.

By the end of the nineteenth century, the colonial economy in Latin America had gradually given way to newly independent capitalist economies and the gradual expansion of wage labor relations, neither of which substantially altered the brutal working and living conditions of coffee workers and small farmers. Most of Latin America had gained formal independence from Spain and Portugal in the 1820s, giving way to formally independent nations run by elite classes who often ruled through a mixture of shame elections (that excluded the vast majority from voting) and political violence against the masses and electoral opponents. Hundreds of years of tumultuous expansion of a highly uneven world system had laid deep roots, and dominant patterns around the social relations of production had been forged, with huge swaths of the best coffee lands owned by powerful agrarian elites. These elites turned from slave labor and debt peonage to hiring landless, low-wage rural workers, while the rest of the coffee lands were divided into tiny parcels owned by marginalized, often

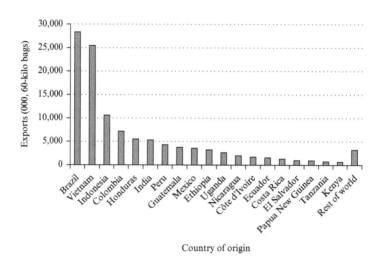

Country of origin

Source: ICO statistical database (http://www.ico.org), accessed November 14, 2013.

Figure 2.1 World's top coffee exporters, 2012.

indigenous, small farmers. Dominant patterns in international trade had also emerged, with Brazil and the rest of Latin America developing and maintaining their status as the dominant coffee growing region in the world to this day.

Deep rooted, however, does not mean static or unalterable, and the world coffee economy was and continues to be dynamic and ever changing. In the chapters that follow, we will see how Colombia emerged as one of the largest coffee producers in the first few decades of the twentieth century, several African countries became key players in the coffee industry with the rise of Robusta beans in the 1950s, and Vietnam made a dramatic move to become the world's second largest coffee exporter in the 1990s (see figure 2.1). In recent decades, Brazil and emerging Asian economies have taken their place among the top coffee *consuming* countries in the world, threatening to topple the United States from its long-standing number one spot in the years to come. The social relations under which coffee is grown have also been altered and contested, provoking both violent backlashes from authoritarian states against the poor majority and, in some cases, genuine and meaningful social reform.

State violence and social change

During colonial times, imperial statecraft was conducted through a direct merger of economic and political power, with colonial states working in tandem with large mercantilist companies granted official trade monopolies by the home state in Europe. The decline of colonial power and the expansion of capitalism brought about a break-up in these official monopolies and the emergence of giant, private trading companies. This transition has often been viewed as a receding of the state from the market. The state, however, continued to grow and expand under capitalism, and in coffee nations, agrarian elites often came to exert overwhelming direct and indirect influence on state policy and its military apparatus. The capitalist state continued to have a "territorial logic," which meant protecting the perceived interests of the coffee economy (a key source of tariff revenues, foreign exchange earnings, national income, and political stability), as well as a "capitalist logic," ensuring the separation of the economic and political, and above all enforcing and policing private property. Given the immense inequalities in coffee countries inherited from the past, protecting private property meant working against social reform or wealth redistribution in a context where the vast majority were poor landless workers and small farmers. This frequently led to extreme violence from the state, often supported by imperialist powers seeking to ensure that the rules of the capitalist economy were preserved at all costs.

Central America is perhaps one of the most telling examples

of the tragic mixture of coffee and state violence in the twentieth century. The legacy of colonialism had left Central American nations with highly unequal distributions of land and wealth, an extreme degree of dependence on the export of coffee and bananas, and a firmly entrenched agrarian elite interested in reproducing and intensifying these patterns. The outcome was pervasive social class conflict, authoritarian violence, rebellion, genocide, and civil war.

The modern history of El Salvador, for example, has been rife with state violence, rooted in the forced displacement of thousands of peasants, beginning in the 1870s, by the state and agrarian elite, much of it carried out to create large coffee estates. The displaced peasants became vulnerable, low-wage workers on the new plantations. When the Great Depression of the 1930s caused large-scale unemployment and general crisis, an indigenous uprising was provoked in 1932 with the goal of regaining possession of displaced land. In response, the state and the agrarian elite unleashed what became known as *La Mantanza* (the Massacre) on the poor indigenous population, ordering the military and the elite's private "White Guards" to slaughter an estimated 15,000–30,000 indigenous people in a matter of weeks. In its wake, El Salvador remained ruled by a series of brutal dictatorships, eventually exploding into civil war from 1980 to 1992. State-orchestrated violence and torture were the norm during the war, all backed by unwavering diplomatic, economic, and military support from the US government. Since the war ended, the state has initiated a series of liberal democratic reforms as part of the peace agreements with rebel groups, along with modest social and agrarian reforms, although inequality and everyday violence remain endemic.

In Guatemala, economic and political inequalities have led to similar historical patterns of authoritarian government and violence. A beacon of hope did emerge during the "Ten Years of Spring," when Guatemala experienced two democratically elected, socially reformist governments from 1944 to 1954. Given that 72 percent of the country's agricultural land was owned by just over 2 percent of the farms, and the substantial majority of the population was rural, meaningful social reform necessitated some land redistribution. When president Jacobo Arbenz attempted such redistribution in 1952, his presidency was brought to an end by a CIA-backed military coup two years later. After Arbenz, the reforms were reversed and the country was led for decades by a series of brutal dictatorships, each one receiving political and economic support from the United States. From 1966 to 1990 these dictatorships massacred an estimated 200,000 indigenous people through a campaign of genocide. A peace accord with rebel groups was signed in 1996, and the state has since carried out liberal democratic reforms without touching the country's huge inequalities in wealth and resources. Consequently, Guatemala remains plagued by daily violence and social injustice, and work and living conditions remain poor for the majority. In one reveal-ing instance, a 2000 survey of coffee farms in the country determined that not a single one paid the required minimum wage – the majority did not even pay half – even though it was so low that the government considered it to cover only 40 percent of a person's basic needs.[11]

An important exception to this general trend in Central America, and much of the coffee world, has been Costa Rica, a small country with a population of around 4.7 million people. Due to a socially reformist, state-led development strategy beginning in the 1940s, Costa Rica has developed as one of the most efficient coffee economies in the world. Moreover, many of its citizens enjoy living standards higher than those of most Southern states and comparable to those of many wealthier nations – Costa Rica's life expectancy at birth in 2011, for example, was 79 years, the same as the United

States. With higher living standards have come comparatively low levels of social and political violence, and Costa Rica has stood out among Central American nations for its relative peace and stability. In fact, while it maintains a national police force, its standing army was abolished in 1949 – something few countries in the world can claim – allowing Costa Rica to avoid the high social and economic costs of militarization and military spending.

More can be said about the Costa Rican case than can be delved into here, and indeed its uniqueness has been the subject of much debate and analysis.[12] Of key importance, however, is an understanding of some of the distinctive aspects of its historical evolution. To begin with, during colonial times Costa Rica was a relatively neglected part of the Spanish Empire, due to its lack of gold and silver, and was sparsely settled – although the settlers who did arrive moved quickly to wipe out its comparably small indigenous population. The result was that Costa Rica differed significantly from the rest of Central America in terms of land and labor: whereas the rest of Central America tended toward relatively scarce land and abundant cheap labor (enslaved indigenous and black workers), Costa Rica tended to have fairly abundant land and a lack of cheap labor. As a result, a substantial class of small and medium-sized farmers along with a more highly paid rural workforce emerged in Costa Rica over time. While a powerful coffee elite still also developed, their lack of dominance over land led them to favor monopolizing the processing, marketing, credit, and export of coffee. Politically, this meant that the Costa Rican agrarian elite developed in a manner less tied to authoritarian rule than did other elites in the region, where state-managed violence was often called upon to protect and reproduce intense inequality and discipline a highly exploited labor force.

The centuries' long evolution of Costa Rica's distinct social and political conditions was combined in modern times with a unique set of historical conditions that emerged to push the political and economic elite toward accepting democratic and social reforms. When the Great Depression of the 1930s, followed by World War II, led to a collapse of coffee prices, small and medium farmers turned increasingly toward radical politics that sought to eliminate elite monopolization over processing and credit. Fearing the political power of masses of independent farmers, the coffee elite felt they had no other choice but to support the more moderate reformist Social Democratic Party (Partido Social Demócrata, or PSD), under the leadership of José Figueres Ferrer, which seized control of the Costa Rican state by force in 1948. Once in power, the PSD proved adept at maneuvering between radical popular movements and the agrarian elite, carrying out liberal democratic reforms and forging state institutions capable of delicately balancing territorial and capitalist logics.

Central to Costa Rica's reformist project was the modernization of the coffee sector. The PSD was able to carry out an agrarian reform without the glaring need for land redistribution apparent in other Central American countries. Instead, the state ramped up taxation on coffee revenues, which it then plowed back into new infrastructure, credit, technology, and agricultural extension services in a manner that gained support from the rural elite, small and medium farmers, and rural workers. Simmering tensions between small and medium farmers and the elite was smoothed by the creation of the state-run Coffee Office, which required wealthy processors to offer minimum prices for coffee beans and loans to farmers under favorable terms. These reforms proved highly effective, and Costa Rica moved rapidly from having a relatively inefficient coffee industry in the 1940s to one of the world's most advanced by the 1970s.

Reforms to the coffee sector were accompanied by key

economic, social, and political reforms on a wider scale. Politically, the PSD introduced universal suffrage and a general strengthening of democratic institutions. Economically, an array of new state-run organizations and regulations were created to manage domestic food prices, provide guaranteed prices for farmers growing products other than coffee, and subsidize retail prices for consumers. Socially, the new state imposed higher wages and stronger labor rights for workers along with a variety of publicly provided services, including health care, education, and social assistance. The outcome was the creation of an effective developmental state with a broad popular support base throughout the country, relative stability and peace, and social indicators well beyond its neighbors and most of the coffee world.

Costa Rica's impressive successes have not, of course, been without shortcomings. While not as unequal as its neighbors, social inequality has remained high in Costa Rica and has become much worse since the early 1980s. Paralleling global trends, the state has gradually abandoned support for social reform in favor of neoliberal policies, cutting public spending, privatizing state assets, reducing crop and food supports and subsidized credit, granting corporate tax breaks, and enthusiastically embracing the rise of new export crops – in particular pineapples, which have come to eclipse coffee since the early 2000s as a source of export revenue – that generally provide low-wage, flexible jobs. At the same time, the Costa Rican welfare state still survives and provides high living standards and important lessons for those concerned with combating poverty and inequality in the coffee industry.

While the unique conditions of the Costa Rican experience mean that it cannot merely be replicated, its history does offer two very important general lessons: first, that a central role for the state in managing the modernization of the coffee sector alongside democratic social reforms can provide significant developmental gains; and second, that the state, under the right conditions, *can be made to play this role*. These lessons are especially astounding when one considers Costa Rica's relatively small size, historical dependence on a limited range of exports, and lack of geopolitical and economic power compared to larger Southern states – all of which suggests that *most* states could do a great deal more to protect and defend the interests of poor coffee workers and farmers than they often tend to do.

Constructing coffee consumption

Coffee consumption has been inextricably linked with the dynamics of coffee production, developing through a long historical process with politics and social power as central driving forces. Production and consumption played off one another, with intensified production frequently driving down prices, inspiring more consumption, in turn sparking more production. Throughout the nineteenth century, these dynamics caused general oversupply and downward prices, which often hit rural workers and small farmers hard, while inspiring Northern workers to increase consumption. By the start of the twentieth century, coffee had become a common component of Northern diets, especially in the United States, which had emerged as the world's largest coffee consuming country. Coffee sales then steadily increased throughout the course of the century, with coffee proving to be particularly well suited for capitalist consumption patterns and the rise of a mass consumer society. Coffee was a small luxury affordable to most social classes, providing energy and "pep," and gained popularity as more localized tastes rooted in class, ethnicity, and geography became replaced with more standardized tastes, aimed at mass-produced commodities from faraway places; coffee from halfway around the world could now be sold as a generic product in chain stores with little popular awareness of where it came from or how it was grown.

While coffee does have certain intrinsic qualities that have made it such a popular drink, it is important to keep in mind that so do many other products. Major marketing efforts on the part of Northern coffee companies and Southern producing states, perhaps mixed with a bit of luck, were required to spur coffee's success, which was not inevitable or guaranteed through market dynamics on its own. Coffee's desirable intrinsic qualities stem primarily from caffeine, a stimulant that provides a temporary boost of energy, although it can also make people irritable and cause dehydration, upset stomach, short-term headaches, and short-temperedness. Recent research has suggested that "excessive" coffee drinking of more than 28 cups per week may result in higher death risk for adults under the age of 55.[13] At the same time, coffee is generally considered to have milder effects than competing drugs, such as alcohol, which is connected to major health problems and known to induce aggressive behavior. As a result, over time coffee has emerged as a less socially regulated and more broadly acceptable stimulant than other, harder drugs.

Coffee has still had many competitors, however, in the "mild" drug category, and coffee promoters have played a key role historically in coffee's international prominence and sales growth, using their economic weight to expand mass consumption by funding expensive marketing campaigns aimed at normalizing coffee addiction and popularizing daily coffee rituals. The twentieth century marked the rise of modern corporate marketing, with ever more expensive advertising campaigns utilized to drive consumer preferences and expand consumption to previously unimagined, and in many ways ecologically unsustainable, levels. The coffee industry, especially in the United States, was a major player in the marketing race, symbolized by the decision of General Foods to spend an unprecedented $2.5 million advertising Maxwell House coffee in 1949. This sparked the rapid escalating of marketing campaigns as corporate competitors rallied with their own massive advertising efforts. Today, the global coffee industry is dominated by giant transnational companies that spend tens of millions of dollars each year on major marketing efforts and hugely expensive branding campaigns.[14]

Alongside the work of Northern-based corporations, coffee states also played significant roles in expanding coffee consumption through economic statecraft, not just through managing increased production but also through direct involvement in marketing campaigns of their own. One particularly famous campaign has been a long-standing one since the late 1950s to promote consumption of Colombian coffee through the fictional character of "Juan Valdez," a campaign funded by Colombia's quasi-state agency, the National Federation of Coffee Farmers (Federación Nacional de Cafeteros, or FNC). Even more famous is the very idea of a "coffee break," which did not emerge spontaneously out of popular culture, but rather was created and promoted by a major advertising campaign in the United States launched in 1952 by the Pan American Coffee Bureau – an organization formed and funded by a handful of Latin American coffee states. From the 1930s to the 1960s, the Pan American Coffee Bureau exerted considerable effort in promoting and expanding coffee consumption. Not all of their campaigns were as successful as the coffee break. In the 1960s, the Bureau launched a "Mugmates" campaign to win over the youth market by asking adolescents to decorate their personal coffee mugs. The campaign failed, steamrollered by more effective and expensive campaigns from the soft drink industry, which also offers highly caffeinated beverages. Soft drinks emerged as one of coffee's great competitors in the twentieth century, winning the youth market and establishing a norm wherein children generally consume soft drinks while adults drink coffee, even though there is nothing particularly natural or healthier about this outcome.[15] The end results were those of marketing battles, waged by coffee states and coffee companies, which have expended considerable resources to ensure coffee's pre-eminence as a daily commodity.

Conclusion

The brief history of coffee provided here offers only a snapshot of the significant patterns in the long history of coffee's production and trade. Its purpose has been to challenge the history offered by free traders, which generally involves a vague depiction of capitalism as the inevitable outcome of the natural human desire to "free" trade and participate in a market economy. In contrast, the expansion of the world system and the emergence of the global coffee industry entailed a conflict-ridden and contradictory history, one in which massive amounts of political and military force were employed by dominant classes to forcibly create a specific set of social relations around production and trade that were far from free. These social relations, then, invariably provided the greatest benefits to the landed elites, slave owners, colonial merchants, and agrarian capitalists who steered their evolution, while reproducing poverty, inequality, vulnerability, violence, and political and economic marginalization for the vast majority.

Statecraft has played a key role in the history of coffee, in particular with the growing involvement in coffee production and trade by colonial states in the eighteenth and nineteenth centuries, gradually eclipsed by modern capitalist states. These states did not merely determine outcomes in a "top-down" manner, but rather had to manage territorial and capital logics under constant pressure "from below." The result was a perpetual "tug of war" between dominant and subordinate groups that forged the basis of a fluid, continually changing "hegemonic consensus" ultimately "skewed" in the interest of dominant groups.[16] History was made through this tug of war, often with painful or incomplete results: Haitian slaves freed themselves from the bondage of the French Empire, although they still pay the price for their rebellion to this day; Salvadorians and Guatemalans, through rebellion, civil war, and surviving genocide, extracted important, yet still inadequate, liberal concessions from oppressive states; and Costa Ricans were able to attain a reformist compromise through the state, with substantial social welfare benefits.

Contemporary patterns of trade and social relations of production have deep roots in coffee's historical legacy. Consequently, unequal distribution of land and resources cannot be dealt with merely by imploring coffee farmers to adopt expensive new technologies and diversify (into what, and at what cost?), or by recommending poor coffee workers to seek alternative jobs elsewhere (unless these can be proven to exist, under better conditions). Similarly, the historically determined, highly vulnerable and dependent position of weaker coffee economies in the international division of labor cannot be altered by beseeching policy makers to diversify and enhance their country's comparative advantage, unless it can be demonstrated how to do this in a highly competitive global economy dominated by larger, richer, and more powerful states. Historically determined patterns of inequality and injustice ultimately require redress from the very states that have been central in creating them. States have seldom stepped up to play this role, least of all as a collective group, but they can be compelled to do so under the right conditions, as chapter 3 reveals.

CHAPTER THREE

Pro-poor regulation

Perhaps one of the most powerfully held notions of the free trade package is the belief that all attempts to regulate international markets have been total failures. This notion is generally accepted without rigorous historical analysis and is frequently employed to silence or dismiss alternative perspectives on trade, development, and the state. In defense of alternative perspectives, and to gain a greater understanding of the power of coffee statecraft, this chapter will reassert the importance and current relevance of the complex history of state-managed trading initiatives that emerged throughout the twentieth century. A central idea in the analysis that follows is that *all* markets are managed by states. This makes the popular distinction between "regulated" and "unregulated" markets difficult to use and in many ways misleading. One way to address this is to make a distinction between the ways in which states regulate markets every day, driven by their territorial and capitalist logics and often without regard for the social and environmental impacts of their actions, and "social" regulation, by which is meant state action conducted in the interests of maximizing welfare for society overall, and in particular with a "pro-poor" bias.[1]

State-managed attempts at social regulation were particularly popular and pervasive in the decades following World War II, from the 1940s to the 1970s, an era often referred to as the age of "embedded liberalism." During this time, a rapid expansion of global trade and investment occurred alongside the emergence of an array of socially interventionist states, from Keynesian welfare states in the North, to "developmental" states in the South, to a complex range of "Communist" and socialist states. Embedded liberalism under these political conditions involved a degree of liberalized trade combined with national controls on capital flows and investment and a range of international market mechanisms designed to manage supply and prices in the interest of poorer states, farmers, and workers. Such international market mechanisms frequently emerged out of strategic statecraft, driven by the desire of Southern economic and political elites to protect or enhance agro-export and extractive industries while restraining rural protest and revolt among exploited and vulnerable rural populations. The mechanisms also developed as part of a growing internationalist movement, during a time of decolonization in much of Africa, Asia, and the Caribbean, and part and parcel of a growing movement demanding fairer global trade and aid rules at international forums like the United Nations.

Collective state-managed social regulation was especially common for primary commodities (such as coffee, tea, sugar, cotton, tin, rubber, wool, and zinc) that tended to have particularly volatile price patterns. While the exact natures of the many mechanisms put in place in the post-war era often differed, they generally were premised on regulations to restrict or manage supply to push up prices on international markets. The degree of success or failure varied widely, with several collapsing with limited or no apparent gains, others attaining mixed results, and still others achieving much greater success, funneling increased income into the hands of Southern farmers and workers and smoothing market volatility and uncertainty. One of the most successful of these agreements was the ICA, a quota system that existed from 1963 to 1989 involving all major coffee producing and consuming countries, designed to stabilize and increase coffee prices by holding a certain amount of coffee off the global market. While generally ignored or dismissed by free traders, the history of the ICA reveals that it provided higher incomes for coffee farmers in need and greater "social efficiency" than the "free trade" coffee regime that has followed in its wake. Despite its shortcomings, the ICA demonstrated both the ability of states to act collectively to attain broader social ends and the benefits of doing so.

Coffee crises and the lead up to the ICA

As discussed in chapter 1, the coffee rollercoaster, its boom and bust cycle, has always been a central feature of the coffee industry due to the specific nature of coffee production and trade. For centuries, coffee farmers have been vulnerable to the swings of the boom and bust cycle. This vulnerability, however, was greatly intensified in the nineteenth century as farmers turned increasingly away from a more diversified mixture of crops, aimed at both local use and global consumption, and toward monocrop production exclusively for sale on the market. Good crop years could bring in solid income for some, and for the largest planters substantial wealth, but bad crop years could be devastating, not just for farmers and workers but for entire national economies. This was driven home beginning in 1896 when the price of green beans on the New York exchange dropped suddenly and substantially in response to boosted production in Brazil. Prices remained depressed for over a decade, compelling major coffee producing and consuming countries to initiate a series of formal meetings to discuss possible ways to address the crisis, including the prospects of a quota system allowing states to control prices through coordinated international action.[2]

In the end, the meetings did not result in agreement on

anything substantial, including the quota system. Such an outcome, however, was not satisfactory for Brazil and especially its major coffee state, São Paulo. On the one hand, declining coffee revenues threaten both the state's territorial logic (with coffee providing a key source of state revenue) and capital logic (with coffee exports bound up with the interests of a powerful agrarian elite). On the other hand, Brazil was uniquely situated to respond to the crisis, given that it accounted for over half the world's coffee beans and possessed significant economic and political weight that could be translated into effective statecraft. Consequently, in 1906 the state government of São Paulo initiated a "valorization" scheme, forming a partnership with Northern banks and coffee merchants, which began purchasing Brazilian Arabica beans to keep them off the market and raise prices. Gradually, the valorization began to have an effect on global prices, which reached 14 cents per pound in 1911 – close to where it had been in 1896. Brazil's success met with fierce resistance in the United States, which accused Brazil of unfairly cornering the market at the expense of US consumers. Under intense pressure, Brazil agreed to sell all of its stored coffee in 1912.

To Brazilian coffee representatives and state officials, the valorization scheme had been a success, revealing the relative effectiveness of coffee statecraft in attaining higher prices.[3] These higher prices also encouraged plantation owners to step up production and plant new trees, which invariably set the stage for another glut and coffee bust. When World War I (1914–18) led to a rapid drop in global demand and prices, Brazil stepped in again. This time, the federal government held the reins, heading the country's second valorization scheme with the purchase of nearly three million bags of beans in 1917. Once again, the valorization was a success. The reopening of core markets when the war ended, combined with a Brazilian frost, initiated a major boom, and the

Brazilian state was able to sell off its stored beans at a significant profit.

While Brazil's unilateral valorization was proving beneficial for its own coffee industry, it was also driving higher prices for coffee globally, which had the effect of encouraging new competitors. This eventually watered down Brazil's unchallenged dominance of the industry. Central America and Colombia, in particular, responded to higher prices by massively increasing their own coffee production, making huge inroads into global markets and dragging prices down by the 1920s. In response, once again, Brazil conducted a third valorization scheme, purchasing millions of bags of its own coffee in 1921. A few years later, Brazil sold the beans at a substantial profit, sparking major protests from the US government and US coffee roasters. Brazil resisted these protests – pointing out that the United States took similar measures for cotton, wheat, tobacco, and other products – and proceeded to carry out a fourth valorization scheme in 1926.

This time, Brazil's luck had run out. On October 29, 1929, the US stock market crashed, initiating the Great Depression, and leaving Brazil holding tens of millions of bags of coffee at a time of plunging prices with no immediate end in sight. Desperate, Brazil banned new coffee planting and burnt millions of bags of coffee in hopes of bolstering prices by reducing available supply, to no avail. On a global scale, the Depression had immense consequences for small and medium farmers, causing mass bankruptcy and social chaos, and for rural workers, who were either laid off or saw their wages cut and poor working conditions intensified. These outcomes led to social unrest, which threatened the territorial and capital logics of coffee-dependent states. In response, many states responded individually, often through terror and violence unleashed on the masses of rural poor by vicious dictatorships, as discussed in chapter 2. The Depression also had the effect, however, of giving a boost to serious talks about the possibility of coffee states working together to address price instability through collective action.

In 1936, Latin American coffee countries met to discuss a collective response to spiraling coffee prices. The meeting was held at the urging of Brazil, which had grown increasingly aware of the limitations of unilateral valorization and progressively frustrated at other coffee countries "free riding" on its efforts. Brazil's valorizations had effectively propped up global prices in the short term, but also encouraged new competitors to enter the market in the longer term. The new producers benefited from the higher prices while the Brazilian state accepted the burden of the risk – all the while sowing the seeds for an eventual glut caused by global overproduction and another bust in prices. Consequently, Brazil pushed hard for a collective resolution at the meeting, which resulted in an agreement to fund a Pan American Coffee Bureau to promote coffee consumption (discussed in chapter 2) as well as a bilateral arrangement between Brazil and Colombia, now the world's second largest coffee exporter, to sell their beans with set minimum prices. Brazil also continued to conduct unilateral coffee statecraft, holding 70 percent of its beans off the market in 1937 and burning over 17 million bags of coffee, at a time when world consumption was only around 26 million bags.

Initial efforts at collective action met with little success. Faced with social unrest and desperate to raise small coffee farmers' income, Colombia broke the agreement with Brazil within a year, selling its beans below the fixed limit to encourage exports. Brazil responded forcefully, calling for another conference of Latin American coffee producers and threatening to abandon its unilateral valorization efforts. When other countries attempted to call Brazil's bluff, refusing to come to terms with a new agreement in the belief that Brazilian state

officials were unable or unwilling to truly abandon valorization, Brazil proved them wrong. It abandoned efforts to hold
its beans off the market and even began to encourage Brazilian
exports with a new coffee tax reduction. Exports boomed,
giving an initial boost to the industry, until prices rapidly collapsed. Prices for Brazilian Arabicas fell as low as 6.5 cents
per pound, resulting in a situation of increased exports but
declining income; according to journalist Mark Pendergrast,
"In 1938 Brazil exported 300 million pounds more coffee to
the United States than the previous year – but received $3.15
million *less* for the total than in 1937."[4]

It was not until World War II (1939–45) that Brazil was
able to attain a multilateral agreement regulating international coffee prices. The war greatly intensified fears of Latin
American coffee countries about continued prices by closing
off European markets. The war also had the unique effect
of pushing the United States into a position of supporting
coffee price regulation. The United States had long protested
against Brazil's valorization schemes and had been an equally
staunch opponent of any proposed collective mechanisms
that could hurt US corporate coffee roasters and erode the
real purchasing power of US consumers. World War II, however, raised the prospect that Latin American coffee countries,
especially Brazil, could be driven into the Nazi or Communist
camps in response to the social and political chaos caused by
the poor state of the coffee market. As a result, the United
States threw its support behind an Inter-American Coffee
Agreement (IACA), signed in 1940, under which it agreed
to a quota of 15.9 million bags for all coffee allowed into the
US market – close to one million over the estimated total US
consumption at that time. This quota was then divided among
Latin American coffee exporters, with Brazil and Colombia
receiving the lion's share: 60 and 20 percent of the quota
respectively. Only around 2 percent of the quota was put aside

for all of Asia and Africa, which were minor coffee exporters at the time. As a result of the agreement, coffee prices almost doubled by the end of 1941, causing some fear among US state officials, who unilaterally increased the quota by 20 percent and froze prices at 13.38 cents per pound from 1941 to 1946.

When the war ended, coffee prices boomed, and the IACA expired in 1948 without renewal or political contention. European markets were reopened and US demand was growing fast, due significantly to the fact that instant coffee had been part of the standard US army food ration during the war. By 1949 the price of Brazilian Arabicas had reached 80 cents per pound – compared to 6.5 cents a decade earlier. High prices encouraged coffee expansion in Brazil, which, combined with innovations in fertilizers, pesticides, and planting strategies, resulted in a major boom in production. High prices also inspired the rapid emergence of new competitors, especially those in Africa who emerged in the 1950s as the dominant exporters of *Robusta* beans.

Over 80 percent of the beans exported from Africa were Robustas, which vary from Arabicas in several ways. Robusta beans take only two years to produce their first harvest (compared to three to five years for Arabica), yield more beans per tree, are more resistant to disease, and grow at lower altitudes in warmer regions than Arabicas. Offsetting these benefits, Robusta beans contain around 50 percent more caffeine than Arabicas and as a result have generally been considered to produce a harsh taste. This historically placed significant limits on Robusta's desirability as an export crop. Matters changed dramatically in the 1950s with the rise of instant coffee, in which bean quality was much less relevant, and with new processing methods adopted by major roasters in the North, who could now blend cheaper Robustas in with Arabica beans. By the end of the 1950s, Robusta beans accounted for nearly a quarter of world coffee exports. Today, over 38 percent of the world's coffee beans are Robusta and are grown through-out Africa as well as by major coffee exporters like Vietnam, Indonesia, India, Brazil, and Ecuador.

The rise of African coffee countries occurred not just in response to price signals, but under the direction of a vari-ety of state coffee agencies that emerged in the late colonial period in the 1950s and after independence. These agencies generally functioned with a state-granted monopoly of coffee exports and provided varying degrees of research and develop-ment, extension services, and credit to growers. They brought in revenues to the state by selling at world market prices beans purchased at lower internal prices. In many instances, this additional revenue was squandered by corrupt or inefficient states. At the same time, despite the difficulties, African coffee agencies did provide some services and market protection to farmers, and in general terms oversaw a major expansion of domestic coffee industries. From the 1950s to the 1970s, African countries led by Côte d'Ivoire, Angola, Cameroon, Burundi, and Tanzania became major exporters of Robusta beans to markets in the colonial and former colonial countries of Belgium, France, Portugal, and the UK. Several African countries, in particular Ethiopia, Kenya, and Rwanda, also emerged as significant exporters of higher-quality Arabica beans. By 1970, African countries had risen from being minor coffee exporters to accounting for a combined total of 32 per-cent of the world's coffee exports.[5]

As has often been the case in the coffee world, the rapid rise of major new exporters caused crises in the global market. The boom of African exports initiated a major glut, which in turn led to new attempts to develop international agreements to collectively manage coffee prices. In 1958, Latin American countries approved their own coffee agreement, with all par-ticipants agreeing to withhold varied amounts of their crop from the market, including 40 percent for Brazil and 15 percent for Colombia. The next year, the same countries agreed to try a one-year quota system, this time with African coffee countries involved, in which everyone committed to export 10 percent less than their best year in the previous decade. While interesting enough on paper, the agreement had only lukewarm support in many of the countries involved and its terms were widely broken, even though it was renewed for a second year. All the while, overproduction continued to wreak havoc on coffee prices, with supply far outreaching demand – by the end of the 1960s, global production had reached 50 million bags compared to only 38 million bags of global consumption.[6]

The nature of the game finally changed in 1962, when formal talks conducted under the aegis of the United Nations successfully led to the 1963 ICA. This multilateral agreement involved the world's most significant coffee producing and consuming countries. Unchecked overproduction and declining prices had reached severe levels, posing major threats to the territorial and capital logics of coffee producing states. Whether large or small, coffee-dependent countries faced economic and social crises, pushing millions of farmers and rural workers toward protest, rebellion, and in many instances radical or revolutionary politics against unequal landownership, terrible wages and working conditions, and the capitalist economy in general. The inability of the state to protect prices and income, combined with declining state revenue from coffee export tariffs and depleted foreign exchange earnings, threatened the overall stability of several coffee states; they simply could no longer afford to sit on the sidelines or strategically avoid meaningful collective action of the sort long pushed for by Brazil and, by this time, Colombia as well.[7]

In the North, support for the ICA among consuming countries harkened back to earlier agreement to regulate coffee prices during World War II for fear that Latin American countries would be driven into the Communist or Nazi camps. The United States in particular was driven by its own capitalist logic, and the fear that low prices would radicalize coffee countries and lead them to adopting socialist, Communist, or economic nationalist development models at odds with the world capitalist system and the interests of Northern capitalists. The signing of the ICA occurred in the wake of both the Cuban Revolution in 1959, which intensified US fears of the spread of Communism, and US President John F. Kennedy's launching of the "Alliance for Progress" in 1961, designed to counter this spread with increased aid and economic cooperation to the South. US government support for the ICA was also combined with growing support from large US roasters. These corporations had historically seen price controls as a threat to their profit margins. Intense pressure from Brazil and Colombia, however, which typically offered the largest roasters special deals for very large contracts, convinced the corporations of the necessity of supporting the ICA and its possible benefits, including leveling both the highs and lows of the coffee cycle and encouraging a stable and reliable supply of beans.

Taming the coffee rollercoaster

The first ICA was a five-year agreement, lasting from 1963 to 1968, although the US Congress delayed ratification until 1965 for fear of the impact of higher coffee prices. Signed by all of the world's major coffee consuming and producing countries, the ICA was a binding agreement managed by the ICO that limited all green bean exports to North America and Western Europe to a quota of just under 47 million bags. Of this quota, over 18 million bags were allocated to Brazil and over 6 million to Colombia, with the remainder divided among other coffee countries on the basis of each one's previous export years. Approval of quota allocations and rules were determined through a voting system in which each group of countries, importing and exporting, were allocated 1,000 votes, and decisions had to be ratified by a two-thirds approval from each group. As the votes were based roughly on the overall size of exports or imports, Brazil and the United States were each granted 400 votes, giving each country veto power over any quota decisions. In addition, the ICA included certificate of origin requirements for all shipments; a quota exemption for countries with low coffee consumption such as Japan, China, and the USSR; and a stipulation of 90 days' notice required from any country opting to withdraw from the agreement.

During the years of the ICA, several challenges emerged to its operation and effectiveness, some of which persisted whereas others were more or less successfully managed. The issue of "tourist coffees," whereby beans were sold to countries exempt from the ICA at lower prices and then re-exported into participating countries to avoid the quota barriers, emerged as a persistent challenge to the ICA system. Another major issue was overproduction, which continued more or less apace, with world production in 1966 exceeding demand by a surplus of 87 million bags. And yet, despite this excess supply, new technologies, such as fertilizers and hybrid plants, were constantly increasing the productivity of the mostly wealthy planters that could afford them. The result was an irrational system wherein countless labor hours and resources were exerted growing millions of surplus coffee bags that no one wanted. In response, Brazil set about bulldozing and burning millions of older coffee trees, but overproduction remained a global problem for many decades to come.

Coffee states were more successful in negotiating new arrangements to deal with disputes about quota sizes and prices. In particular, countries with traditionally smaller volumes of exports complained that quota limits did not take into account their increased production, and countries with higher-quality coffee beans argued that the quotas unfairly restricted their prices and market growth. To allow for greater flexibility, the ICA quotas were revised with new target price ranges: quota increases were automatically triggered if world market prices rose above the target price range, and similarly automatically decreased if market prices dropped below the price range. In addition, the price range was now determined by a "selectivity principle" offering different prices for different qualities of coffee: Robustas (primarily from Africa and Indonesia at the time); unwashed Arabicas (dominated by Brazil); Colombian milds (Colombia and Kenya); and "other milds" (Central America).

The ICA was determined to have met the needs of coffee states enough to be renewed from 1968 to 1972. During the second ICA, disputes between exporting and importing states began to emerge over quotas and prices. In 1969, the price of Brazilian beans fell to 35 cents, just above the target price of 34 cents. Fearing the downward trend, nine major coffee exporters from Latin America and Africa met in Geneva, forming what became known as the "Geneva Group," and formally requested lower quotas to boost prices. When prices rose abruptly the following year, up to over 50 cents, due to a frost and a drought in major coffee regions in Brazil, the simmering dispute was temporarily sidelined, and producing countries even agreed to raise the quotas to allay fears from the United States over the price spike. In 1971, matters heated up again when the United States abandoned dollar–gold convertibility and devalued the US dollar in response to the rising costs of the Vietnam War and a declining position in world trade. As coffee was traded internationally in US dollars, the effect of the devaluation was to dramatically reduce the real prices of coffee. When the United States rejected a request from coffee countries to lower ICA quotas to compensate for the devaluation, the members of the Geneva Group opted for a different form of coffee statecraft, deciding to collectively hold their beans off the market beyond the agreed terms of the ICA. This successfully raised prices, as well as the ire of the United States. With tensions mounting, the ICA expired without renewal at the end of 1972. The trading of futures contracts on the New York exchange rapidly took off, with speculators anticipating a return to more extreme price swings.

Despite mounting opposition among US politicians and policy makers to price regulation, the United States was compelled to sign a new ICA only four years later, in 1976. This decision was driven to a large extent by a rapid increase in coffee prices. From 1975 to 1977, the coffee composite indicator price more than tripled, soaring from 63 cents to $2.29 per pound. The price boom was driven by an unprecedented snowfall in Brazil in 1975, the "Black Frost," which destroyed almost the entire national harvest for the year, as well as a combination of unrelated political and environmental events throughout the coffee world, from civil wars and labor unrest, to flooding in Colombia, an earthquake in Guatemala, and a coffee leaf rust outbreak in Nicaragua. Under these conditions, prices skyrocketed, leading to consumer protests and congressional hearings in the United States, all of which compelled the US government to sign back on to the ICA in hopes of finding a way to somewhat stabilize prices. Coffee states, despite enjoying high prices, were ready and willing to participate in a new ICA, anticipating the likelihood of downward price shifts in the near future.

This is precisely what happened, and by 1978 the indicator price had fallen down to $1.55. This price seems high by historical standards and was above the ICA quota trigger price, but in real terms it was low due to the devalued US dollar. Coffee countries requested a higher trigger price from the

United States but were rebuffed. As a result, a group of coffee countries once again turned toward collective coffee statecraft to push for a renegotiation of the terms of the ICA. This time, major Latin American producers met in Bogotá, Colombia, and formed the unofficial "Bogotá Group," with the goal of using $140 million in funds to speculate on the New York coffee exchange in a manner designed to drive prices up. The Group met with success, driving prices up by several cents or more, and even creating its own trading company, Pancafe Productores de Café SA, in 1980 to formally continue with its activities. At this point, the United States stepped in and agreed to a new ICA with higher target prices in exchange for the disbanding of the trading company.

The ICA continued throughout the 1980s and was renewed in 1983 and 1987. This created higher and more stable prices than was typically the case in the coffee world, although prices continued to fluctuate greatly throughout the decade, from as high as $1.70 to as low as $1.07 per pound. Brazil and Colombia remained staunch supporters of the ICA, but agreement among other coffee countries was breaking down, in particular over the view that the ICA was unfairly limiting the exports of the most efficient and higher-quality producers. Costa Rica, for example, had developed into one of the most efficient coffee economies in the world, producing higher-quality "milds," but due to the limits of the quota system regularly had to sell 40 percent of its crop at significantly lower prices on markets outside the ICA system – which were then often re-exported back to major markets as "tourist coffees." Support for the ICA in the United States was also breaking down. Major US roasters increasingly viewed the ICA as a barrier to corporate profitability, restricting their access to higher-quality beans and preventing them from benefiting from global overproduction, which would drive lower green bean prices and higher profit margins if unchained from the quota system. The US state, for its part, had always been a reluctant supporter of the ICA due to Cold War politics that compelled it to accept price regulation against its growing dedication to neoliberal reforms and "free trade." The fall of soviet-style Communism in Russia in 1989, heralding the end of the Cold War, freed US negotiators and set the stage for the final curtain on the ICA.

In 1989, the United States took a hard line on negotiations for new quarterly quotas for the ICA. Brazil and Colombia fought to keep the agreement, supported by most African producers and most European importers. By this time, however, significant opposition had developed among smaller, higher-quality Arabica exporters, including Costa Rica, Dominican Republic, Ecuador, El Salvador, Guatemala, Honduras, India, Mexico, Nicaragua, Papua New Guinea, and Peru. Enough votes could not be mustered to approve new quotas and the ICA was suspended. The United States then formally withdrew from the ICO four years later in 1993.

The end of the ICA initiated a major drop in prices and intensified market volatility. Long-standing global overproduction, now unleashed, led to a swamping of global markets, made worse by the unchecked, rapid growth of new leading coffee exporters, in particular Vietnam (discussed in chapter 4). The coffee indicator price fell from $1.15 per pound in 1988 to 54 cents in 1992, recovered somewhat from 1994 to 1998, and then collapsed, dropping to 45 cents by 2002 – the lowest price in 30 years and, according to Oxfam International, probably the lowest *real value* in over 100 years, taking into account inflation.[8] Extreme market volatility was further intensified by a flurry of new speculative activity. The total volume of futures contracts traded on the New York exchange in 1994 reached nearly 10 times the physical volume sold, so that, states Talbot, "by the mid-1990s, the vast majority of trades made on the coffee futures markets were made for purely speculative purposes."[9]

The crisis in prices translated directly into social and political crises throughout the coffee world, sparking collapsing incomes, unemployment, bankruptcy, migration, hunger, and increased poverty for thousands of small and medium farmers and coffee workers. African coffee countries were among the worst hit by the crisis, as it came after two decades of thorough "free trade" reforms that had privatized and liberalized the coffee sector, dismantling most state coffee boards and leaving the public sector weak and unable to provide urgent support for farmers or protection for the industry overall. From a high of 32 percent in the 1970s, Africa's share of world coffee exports dropped to around 10 percent, where it remains today. Côte d'Ivoire, once Africa's leading coffee country, experienced a 69 percent drop in coffee production from 2000 to 2009, with devastating impacts on farmers and workers. Amid the worst years of the crisis, Northern transnational roasters continued to make huge profits, taking full advantage of the gap between collapsed green bean prices and much higher retail prices for roasted beans in the North.[10]

The profound impact of the global coffee crisis threatened the territorial and capital logics of many coffee states – sparking social unrest and declining state revenues, and damaging domestic coffee industries – drawing many of them back toward supporting renewed price regulation, including several countries that had actively opposed the ICA in 1989. A political consensus for new quotas, however, could not be reached, for a variety of reasons, but in particular due to fierce and unwavering opposition from the United States. Coffee states were able to pressure for new "International Coffee Agreements," in 2001 and 2007, but they lacked any of the substantial regulatory mechanisms of the past. The 2001 agreement was entirely market-friendly, offering various programs aimed at supporting or promoting coffee training, access to information, marketing, technology transfer, enhanced bean quality, "sustainable" coffee growing, and increased coffee consumption. For all of these initiatives, the ICO had at its disposal $45.2 million, equivalent to less than 1 percent of the combined annual coffee sales of the world's four largest coffee companies.[11] The 2007 agreement (which did not go into effect until 2011) followed the same lines, along with offering a reorganization of the ICO's operating structure in response to declining funds and capacity.[12] Pleased with the market-friendly nature of these new agreements, the United States was persuaded to rejoin the ICO in 2004 after an 11-year absence.

After the darkest days had passed, coffee prices eventually began to recover, climbing over $1 per pound in 2007. Prices then leapt up to $1.95 per pound in 2011, reaching the highest levels in over 30 years. The price boom was driven by the combination of a poor harvest in Colombia, increasing demand for coffee among the growing middle classes in Brazil and throughout Asia, and a spike in speculation on commodity futures as investors scrambled for alternatives to stocks and bonds in the wake of the global financial crisis beginning in 2008. Before the celebration over the new price highs could really get underway, however, prices had already begun to fall. By the end of 2013, coffee indicator prices had dropped to around $1, the lowest prices in six years, in response to slowing economic growth in major coffee markets. Alarmed by these trends, and with effective collective state action nowhere in sight, Brazil once again turned to its old tool of economic statecraft, unilateral valorization, announcing in 2013 plans to purchase up to three million bags of Brazilian beans in the attempt to halt declining prices. Prices then began upward movement in 2014, this time in response to a major drought in Brazil, one of the worst in decades, destroying crop yields throughout the coffee regions.[13]

Assessing the ICA

The quota system of the ICA provided higher and more stable prices than the "free trade" decades that followed it, which meant higher incomes for the world's coffee farmers and workers. If we look at the average annual composite indicator prices for green beans from 1963 to 1989 (ICA years) compared to 1990 to 2011 (post-ICA years), the numbers can appear to be quite close: from 1963 to 1989 the average price was just over 94 cents per pound; from 1990 to 2011 the average was just under 94 cents. This, however, does not tell the full tale. When the ICA began in 1963, prices for that year were only 32 cents. The ICA led to years of steady increase, so that the price never dropped below 32 cents again and, after 1976, the price never dropped below $1 until the last year of the ICA. The post-ICA years, in sharp contrast, began with a

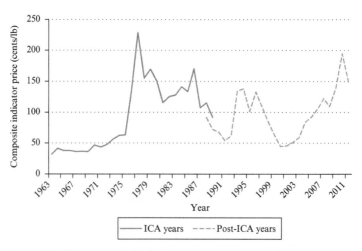

Source: UNCTAD statistical database (http://unctadstat.unctad.org), accessed July 30, 2013.

Figure 3.1 Comparing ICA to post-ICA, free market prices, 1963–2012.

price of 91 cents in 1989. After that, prices swung erratically from extreme lows to extreme highs, going above 91 cents in only 11 out of 22 years, and at some points reaching a low of 45 cents. Thus, whereas the ICA years were characterized by higher annual prices and steady price increases (with the average price in its last decade nearly three times higher than when the ICA began in 1963), the post-ICA years have been characterized by highly volatile price swings that have, over a 20-year average, only just been able to get over the price that the free trade era began with in 1989.

Given this reading of historical prices, free trade economists have generally not been able to oppose the ICA on the basis of it having resulted in lower prices for farmers. Instead, the ICA has been criticized for having artificially propped up prices, preventing farmers from responding to price signals by intensifying production or shifting into other, more viable market activities. This critique, however, fails to soberly assess the position of small farmers in real-world coffee markets. These farmers generally lack proper information on global market trends or the training to interpret these trends. The costs of improving productivity or transitioning into other export crops are usually prohibitive to small farmers. And the existence of other, more feasible market activities, such as decent wage labor off the farm, is assumed, even while the actual options available often involve illegal migration or low-paid, low-skilled work, living on the margins of poor urban shantytowns. The effect of the free trade critique of the ICA, as Louis Lefeber and Thomas Vietorisz have noted in their general critique of conventional economics, is to ignore the social impacts of real-world market conditions in favor of a narrow vision of economic efficiency that "may not be conducive to the enhancement of social welfare, and may even lead to the opposite." Instead, Lefeber and Vietorisz argue for policies based on "social efficiency rooted in concrete social

problems" that consider the broader needs of "the social, economic, political and cultural system as a whole."[14]

From the perspective of social efficiency, the ICA quota
system was superior to the post-ICA years. Through collective
action, coffee producing and consuming nations were able to
attain higher and more stable prices for coffee farmers and
workers. Importantly, these higher and more stable prices
reached all of the world's 25 million coffee farmer families.
Fair trade coffee certification, by contrast (the subject of chapter 5), also seeks to offer higher and more stable prices and
reaches around 670,000 families – about 3 percent of the
world's total. Moreover, while fair trade prices are guaranteed
to be above conventional ones, they remain tied to a range
defined by conventional prices and are not high in historical
terms. From 1976 to 1989, the regular price of conventional
coffee beans under the ICA system was close to or above what
is today considered the "fair trade" price.[15]

Despite its general benefits, however, there are a number
of important criticisms of the ICA that cannot be overlooked.
It was continually plagued by issues around "tourist coffees"
and overproduction that its members were ultimately unable
to resolve. While the ICA did temper the extreme volatility of
the coffee rollercoaster, it certainly did not eliminate it, and
major price swings continued to pervade the industry. While
the ICA did result in greater coffee income being retained in
the South, it had little impact on how this income was distributed; internal politics within each coffee country played
the determining role in this regard. Countries that pursued
statecraft that involved a degree of redistribution of economic
and political power into the hands of smaller farmers and
workers, such as Costa Rica, were able to attain broader developmental gains than highly unequal coffee countries, such as
El Salvador, Guatemala, and Brazil, where much of the additional income went into the pockets of the wealthy.

The same can be said about the negative ecological impacts of the modernization of coffee production, which was never on the radar of the ICA during the quota years, even while it has emerged as a key issue at the ICO today. With the introduction of the "Green Revolution" in the 1970s, chemical-intensive, monocrop production under full-sun conditions grew substantially, covering 30 to 40 percent of Latin America's coffee lands by the start of the new millennium. The most efficient coffee economies led the charge, with full-sun cultivation taking over 68 percent of Colombia's and 40 percent of Costa Rica's coffee land. The negative ecological impacts have included the destruction of tropical rainforests; declines in the biodiversity of trees, birds, animals, and insects; environmental pollution; and increased soil erosion. The decline of natural predators to coffee pests has left crops increasingly vulnerable to major outbreaks of *la broca* and coffee leaf rust that can devastate entire national coffee industries. Workers and farmers have been exposed to a variety of chemicals – through direct contact and through the pollution of local groundwater – which have been implicated in certain cancers, birth defects, and other illnesses.[16]

Conclusion

In the final analysis, the ICA could not address the deep historical and structural roots of inequality and social and ecological injustice in the coffee industry, nor was it intended to do so. From the start, the mission of the ICA was to prop up prices in a manner that benefited the wealthy coffee elite (who had more coffee to sell and more extra income to make) as much as it did smaller farmers and workers. And yet, the ICA was still more socially efficient and "pro-poor" than the free trade regime that has followed. With the end of the ICA, collapsing prices and dramatic price swings brought about two decades of chaos for thousands of poor farmers and workers while the coffee elite continued to survive under intense price competition, and in many instances thrive. With this in mind, the actual, messy history of the ICA offers two very important lessons for anyone concerned about human development in the coffee industry.

First, the ICA provided higher and more stable prices, allowing for better working and living conditions for millions of coffee farmers and workers. While free trade economists have often objected to this, asserting that the ICA prevented necessary diversification into other economic activities, it must be remembered that better employment does not always exist and other commodities face the same global competitive pressures as coffee. Second, the ICA reveals the power of collective action to achieve its stated aims around higher prices. While collective action was ultimately limited and eroded by wider pressures stemming from a highly competitive and uneven capitalist economy, ceding to these pressures in the post-ICA era has only made things worse for the poor and most vulnerable. This raises questions around whether or not *more*, rather than *less*, collective action is ultimately required to address the root causes of historically derived inequality and the impacts of cutthroat competition. The history of the ICA suggests that collective action is both possible to attain, under the right political conditions, and more socially efficient than all-out global competition.

CHAPTER FOUR

Coffee unleashed?

The collapse of the ICA ushered in two decades of chaotic price swings and crises for coffee farmers and workers. As described in chapter 3, coffee indicator prices bottomed out in the early 1990s, recovered somewhat in the middle of the decade, then reached their darkest period from 1998 to 2002, during which time prices dropped as low as 45 cents per pound. This sparked layoffs, bankruptcy, declining incomes, and migration, and intensified hunger and poverty for tens of thousands of poor coffee families. It was not until 2007 that coffee prices recovered, spiraling upward until 2011. They then declined again, to around $1 per pound by the end of 2013, only to begin to shift upward in 2014 in response to a major drought in Brazil. Free traders and their opponents have offered dramatically different assessments of these turbulent years. To free traders, market turmoil has been part of necessary processes of adjustment, required to throw off the shackles of market "intervention" and get on the right course toward full-fledged market liberalization. To opponents, the years of instability following the collapse of the ICA and culminating in the coffee crisis are frequently evoked to demonstrate the chaotic impact of state withdrawal from the coffee industry.

The latter view paints a generally accurate picture of the key dynamics underpinning the coffee crisis. The end of collective action among coffee states through the mechanisms of the ICA intensified market volatility and fuelled price speculation. After decades of overproduction, the removal of supply quotas unleashed a flood of beans onto the global market. Oversupply was then exacerbated by a production boom in Brazil combined with the unanticipated growth of traditionally minor coffee exporters, led by Vietnam. Since the early 1980s, coffee production in Vietnam had been gathering steam. When prices for Robusta beans temporarily spiked from 1995 to 1998, Vietnamese coffee farmers eagerly ramped up production, responding to what the World Bank has termed a "uniquely favorable set of developments in the world market for coffee."[1] The favorable conditions were then quashed by Vietnam's own unprecedented growth, swamping the coffee market with unanticipated supply and triggering a collapse of global prices. In only two decades, Vietnam had climbed from relative coffee obscurity to the world's second largest exporter, surpassing the export volumes of long-standing second-place leader Colombia. Had the ICA been in place during the 1990s, Vietnam's rapid entrance would likely have been better managed and its tumultuous impact – for farmers internationally and in Vietnam – significantly dampened.

While much of the coffee crisis can be attributed to a lack of collective state action to properly manage the world coffee economy, however, this should not be confused with a general lack of state *involvement* in the market. While states were no longer working collectively to buttress prices, they did continue to manage the rules and regulations required for the exchange of coffee in a capitalist world system, and individual states continued to seek to gain advantage over others in the interests of economic statecraft. The end of the ICA was not the end of state regulation, but rather a shift from a degree of collective state action to intensified competitive coffee statecraft. It was, in fact, individual coffee statecraft that played a central role in Vietnam's coffee boom and its global ramifications. The centrality of the state to Vietnam's coffee industry was emphasized by a 2004 World Bank report, which stated that the Vietnamese government:

> has for decades been an integral part of the coffee sector's development. Not only have policies and regulations governed the sector, but government has also directly participated in every aspect of the coffee industry. From input and credit markets to production, processing, and marketing, its influence has been all-encompassing. Government is the primary and most influential institution by far, and has created nearly the entire sector's other institutions.[2]

Thus while it is true that Vietnam's rapid entrance into the coffee market played a major role in causing the coffee crisis, its entrance did not occur strictly out of spontaneous market forces of supply and demand. Rather it emerged out of a conscious effort by the Vietnamese state to promote coffee statecraft. Seeking to take advantage of a history of cultivating modest amounts of Robusta beans, which are well suited to the growing conditions of the Central Highlands, and of mounting interest in Robusta among major coffee roasters, who had improved processing technologies allowing them to better blend bitter Robusta beans with Arabicas, the Vietnamese state sank substantial resources into promoting coffee exports. The result was extraordinary export growth for the country, amounting to an annual average of 29 percent from 1981 to 2001. The centrality of the state to this growth serves as a telling case revealing the continued and unavoidable role of statecraft in shaping coffee markets, even in the era officially dominated by "free trade." This suggests the need to rethink dominant assumptions about the relationship between the coffee market and the state in the era of globalization.

Colonialism and capitalism in Vietnam

To understand the current role of coffee statecraft in Vietnam, it is necessary to reflect briefly on its historical dimensions, which have significantly impacted the way it has been integrated into the world system. As highlighted in chapter 2, the widespread shift from growing agricultural products primarily for family and local consumption toward growing products predominantly for sale as commodities on global markets does not just emerge spontaneously, but requires the right political and social conditions to bring such a massive reorientation about. In the coffee world, these conditions have historically been created through hundreds of years of colonialism, slavery, and imperialism, leading to the development of an economically integrated world system, in which coffee is one important global commodity among others. True to these global patterns, in Vietnam, the existence of export-oriented coffee farmers did not just appear in response to market forces – millions of peasants deciding of their own volition that they would rather grow coffee for export than continue growing food for their own use – but rather were created through a long historical process managed by states.[3]

Like that of much of the coffee world, Vietnam's integration into the world system, initially as a producer of primary commodities, was driven by European colonialism. From 1887 to 1954, Vietnam was colonized by the French Empire, whose primary goal was to extract cheap commodities from Vietnam through the creation of a plantation economy. During the colonial period, the French devoted their efforts to many commodities, including opium, salt, rice, tea, rubber, pepper, and coal. Coffee was also grown on a relatively small scale and was only of minor interest to the French in Vietnam. To produce these products, land was forcibly taken from its original inhabitants and turned into huge plantations that were granted to

French settlers and local collaborators. The rural majority were either left landless through this process, forced to work as "free" laborers on plantations under terrible conditions to survive, or were compelled into tenant farming on small plots, under highly unjust terms. The Vietnamese were also subject to forced labor, directed and overseen by the colonial state, which frequently involved a couple of days per month working on French plantations or building colonial infrastructure.

The French were eventually forced out of Vietnam in 1954 after a decade-long war and a national liberation struggle led by the Viet Minh, a coalition of Communist and nationalist groups. As the French pulled out, however, the United States moved in. Fearing the popularity of the Communist party, which was in charge of North Vietnam and likely to win unification elections scheduled for 1956, the United States threw its support behind a relatively unpopular South Vietnamese government, leading to the Vietnam War between North and South from 1954 to 1975. To the United States, an officially Communist and nationalist government in Vietnam posed a threat to its territorial logic (the perceived need to "contain" Communist expansion during the Cold War) and its capital logic (the necessity of promoting private property and expanding the process of global market integration initiated by the French). The war ended in 1975 with the Communists politically victorious and Vietnam united as one independent country. The United States had, however, attained victory in defeat, leaving behind a country devastated by decades of conflict that had claimed millions of lives, destroyed infrastructure, and wrecked the national economy.

Under these conditions, despite its formal devotion to "Communism," the Vietnamese government soon turned away from efforts to promote collective ownership and toward the transition to a capitalist economy managed by a state elite. Confronted with economic stagnation and hyperinflation in the post-independence years, and blocked by its own authoritarian structure and resistance to open participation and democratic input, the Communist government initially proved ineffective at economic planning, leading to corruption, fraud, and declining political legitimacy. This was made worse by continued and persistent pressure from the United States and its allies through economic statecraft. The United States kept up a trade embargo for two decades and did not normalize relations with Vietnam until 1995. Other Western allies, along with the World Bank and International Monetary Fund, applied constant pressure on Vietnam to liberalize its economy and privatize state assets, offering bilateral and multilateral loans in exchange for commitments to neoliberal reforms and to the pursuit of export-oriented agriculture to earn foreign exchange to meet debt payments. Throughout the 1980s and 1990s, Vietnam's state elite increasingly warmed to this process, and a capitalist logic became entrenched within state policy. This does not necessarily mean "free trade" – in terms of evenly eliminating protectionist barriers – but rather state policies limiting how and when the state can directly intervene in "economic" matters.[4] This requires formally separating the economic from the political realm through the institution of private property.

The Vietnamese state began a gradual transition toward private property rights in 1981, passing a law permitting individual households on cooperative farms to manage their own plots, if they agreed to provide a specified share of their output to the cooperative and submitted to conditions determining the quantity and nature of the crop. Wider reforms were then initiated in 1986 under the banner of "Doi Moi" ("economic renewal"). This included a broad array of changes aimed at a staged dismantling of collectives, the privatization of state assets and institutions, and the gradual liberalization of markets. In the agriculture sector, Doi Moi entailed intensified efforts to encourage small farmers to switch from growing food for local use to export crops. It also involved new initiatives to deepen and expand semi-private property rights. In 1993, the state passed legislation allowing farmland to be designated to individual households for 20 years for annual crops and 50 years for perennial crops, subject to approval by district-level people's committees. It also allowed the land to be traded, inherited, and used as collateral for loans. This was followed in 1999 by the Enterprise Law, which laid the groundwork for the expansion of the private sector through the intensified privatization of state-owned enterprises (SOEs), which dominated the Vietnamese economy from the 1970s to the 1990s. These processes of privatization, private property development, and further global market integration are ongoing to this day, and form the background for the emergence and growth of the Vietnamese coffee sector.

Building coffee statecraft in Vietnam

Just as the general emergence and development of export-oriented farmers on semi-private land in Vietnam did not appear out of market dynamics on their own, the same is true for the more specific emergence of coffee as the crop of choice for hundreds of thousands of these farmers. The French first introduced Robusta coffee beans to Vietnam in the nineteenth century on a relatively small scale, which was how things remained until the 1970s. At that time, most Vietnamese peasants knew little about coffee cultivation and certainly did not have extensive access to timely information on global market trends. Given this, how did smallholders even weigh the prospects for growing and selling coffee, determining its local processing needs and its marketing and distribution networks? The state took the leadership role in this regard. As Vietnamese policy makers gradually succumbed to a capitalist logic, promoting the expansion of a capitalist economy, they also had to remain attuned to the state's territorial logic, seeking to ensure that the specific needs of the Vietnamese state were met and its interests were defended in competition with other capitalist states.

The two logics dovetailed with each other as the process leading toward the large-scale adoption of coffee unfolded. If the state was going to promote semi-independent farm household production, what would the new farmers grow? This is no simple question with a clear answer. Viable crops had to be determined to promote rural incomes, and by extension state taxation and revenue. Even as the state turned increasingly toward promoting urban, industrial growth, rural areas still had to be kept stable to dampen protest or rebellion and slow the flow of rural-to-urban migration. By the 1970s, Vietnam had emerged as one of the world's largest rice exporters, a position that it still holds today. Economic stability, however, required a degree of diversification and not all regions had land and climate suited to rice production. New export options were also required to meet the needs of the emerging private sector – trading and processing companies needed commodities to trade in, banks needed people to lend to, and domestic industries needed consumers to sell to.

In this context, coffee statecraft emerged as a main goal of the Vietnamese state, centered on the provinces of the Central Highlands, in particular Dak Lak, which had soil and climate conditions well suited for Robusta beans and a history of growing small amounts of coffee. Dak Lak was also strategically significant. During colonial times, it was the location of many large French plantations, and during the Vietnam War it was a main area of conflict, controlled by the US-backed South Vietnamese, who took various measures to compel settlement and increase commodity exports from the region. After the war, the victorious North appropriated the former

French plantations and expanded agricultural production through deforestation, creating hundreds of new state farms and cooperatives. Not all the land used for the new farms was taken from colonial oppressors, and the state also unilaterally seized indigenous land, traditionally held in common. Several New Economic Zones (NEZs) were declared in Dak Lak in the 1970s, and the state initiated a major resettlement program that brought hundreds of thousands of new settlers into the region. The resettlement program flooded the region with groups loyal to the government, most of them ethnic Kinh farmers; as a result of the resettlement, within three decades, Dak Lak was transformed from a province with a minority Kinh population to one in which the Kinh were 70 percent. It also provided the state with an escape valve for funneling growing rural populations from relatively poorer and more densely populated areas in the northern and central coastal provinces.

The new settlers were encouraged to grow export crops, including coffee, tea, and rubber. Initially, these commodities were used in exchanges between the Vietnamese state and Soviet Bloc allies for industrial products and technological support. Coffee emerged as the dominant export crop, gaining increasing attention and support from the state. During the late 1970s, the national state encouraged coffee cultivation by offering preferential credit to growers and traders, coffee export bonuses, and a range of government programs to facilitate land access in the Highlands areas. Provincially, state farms provided a variety of technology and extension services for coffee farmers. As Vietnam began increasingly to tap into global markets in the 1980s, these policies were stepped up, offering existing and potential coffee farmers access to preferential loans, subsidized inputs, low-cost land, extension packages with seedlings and chemical fertilizer, and support for irrigation and intensive farming techniques. The state also imposed various controls on domestic food prices, which had the effect of facilitating urban industrialization (by ensuring cheap food for urban workers) while also encouraging farmers to switch to export crops in hopes of attaining a higher income.

By the mid-1980s, the state had successfully laid the groundwork for a substantial and growing coffee industry in Dak Lak. The state no longer had to directly promote migration to the region, which was spurred on by income and job growth. As part of general economic reforms associated with Doi Moi, coffee farmers were gradually allowed more control over their land and the selling of their products, and the government reduced various restrictions on the import of chemical fertilizers. The state, however, continued to directly and indirectly manage the coffee economy in the interests of statecraft, even during the 1990s when economic reforms were intensified.

One important area where the state has maintained involvement in the coffee industry has been through SOEs. These enterprises have had wide mandates, well beyond those of private firms, providing public services, building infrastructure (such as schools, roads, and health clinics), and offering agronomic and technical support to communities. Throughout the coffee region, state-owned plantations played a key role in providing technology and extension services to all farmers (those within and outside of state farms) from the 1970s to the 1990s. Their numbers have declined since then, due to the gradual dismantling or privatizing of SOEs. In the 2000s, state farms only controlled around 5 percent of Vietnam's coffee lands. They continued, however, to account for 15 percent of the country's coffee production and to exercise considerable influence over the industry.

In other areas, SOEs have retained more influence, with state processing and trading companies accounting for around 40 percent of all Vietnamese coffee exports. The most significant state-owned coffee enterprise is the Vietnam Coffee Corporation (Vinacafe), which was created in 1995, as economic reforms were picking up pace. Vinacafe was designed to take control of a wide range of activities previously run directly by the state, including coffee growing, research, processing, and the provision of credit, fertilizer, and irrigation. It is one of the country's largest SOEs, running dozens of state subsidiaries (40 of which are state farms and 27 of which are coffee processors, traders, and service providers), employing 27,000 people (with an additional 300,000 seasonal workers), and managing substantial industries. Vinacafe owns one of only two instant coffee facilities in Vietnam (the other is owned by a subsidiary of Nestlé), and one of its trading subsidiaries is one of the largest single coffee exporting companies in the world, regularly exporting over three million coffee bags per year, well above the national production of the majority of coffee countries. Through Vinacafe, the Vietnamese state has maintained control of an institution of great economic and geostrategic weight.

The state has also continued to manage the coffee industry through its control of a range of incentive mechanisms, subsidies, and credit. Of these, credit perhaps stands out the most, with Vietnam recognized as having among the most reasonable and extensive credit offerings to its rural sector of any developing country in the world.[5] Most rural credit comes from the state-owned Vietnamese Bank of Agriculture and Rural Development (VBARD), which has around 1,600 rural branches and controls 75 percent of the credit to the country's coffee farmers. For the smallest farmers, there exists a micro-lending institution, the Vietnam Bank for Social Policy (VBSP), which, while nominally independent, is underwritten by the state and funded predominantly from compulsory contributions from state-owned banks. Through its relatively

substantial credit offerings, the state has been able to promote the expansion of coffee production while maintaining a
degree of control over individual farm decisions through the
terms and conditions that accompany loans.

Beyond state enterprises, the Vietnamese state has also
maintained control over direct tools of statecraft in a manner
that has diverged somewhat from the dominant consensus
around neoliberal reforms and "free trade." Whereas most
coffee states in Latin America and Africa in the 1980s and
1990s focused on attaining competitive advantage through
rapid market liberalization and the privatization of state institutions, Vietnam instead pursued a more gradual, piecemeal
liberalization and privatization, combined with a relatively
robust role for state management. In the mid-1990s, when
coffee prices were good, Vietnam mandated that exporters
pay fees to contribute to a government-run price stabilization
fund, designed to offer support to farmers if prices fell below
the costs of production. (In the end, the fund was primarily
used to pay for subsidized credit programs.) When the global
coffee crisis began in 1998, the state eliminated the fees associated with the fund to lessen the burden on coffee exporters;
directed state-run enterprises to store coffee beans in hopes
of reducing supply and boosting prices; and, in 2001, ordered
banks to freeze loan repayments for coffee growers for up to
three years to prevent general default.

The Vietnamese state has also taken a somewhat different path in relation to dominant trends in the coffee world
and beyond at a global scale, taking part in international
forums and initiatives, but with less enthusiasm or limited
involvement. Coffee statecraft on an international level has
increasingly been aimed at aggressively pushing for market
liberalization, through institutions like the World Trade
Organization (WTO), to pry open rapidly growing new coffee
markets, like Russia, and to facilitate the smooth shipment

of coffee beans. Coffee statecraft has also increasingly been aimed at promoting "sustainable coffee," the stated goal of the two new "International Coffee Agreements" in 2001 and 2007 (discussed in chapter 3). This essentially involves efforts to promote quality and various symbolic attributes to gain higher prices through the support of international non-state certification initiatives and consumer movements (such as fair trade, the subject of chapter 5). Overall, Vietnam has been less enthusiastic about these trends, somewhat reluctantly joining the WTO in 2006, 12 years after its creation and after Vietnam had become firmly established as global coffee leader, and preferring to construct its comparative advantage not through "sustainable" coffee, but around chemical-intensive production, beating out competitors by producing huge volumes of coffee beans sold at low prices. By one estimate, only 10 percent of Vietnam's coffee beans can be considered "sustainable," involving some sort of certification, compared to 75 percent in Latin America.[6]

Vietnam's unique form of coffee statecraft set the stage for the rapid growth of its coffee industry in the 1980s and 1990s. As this summary makes clear, it is a misnomer to depict Vietnam's rapid entry into the coffee market, or the global crisis it contributed to, as events that were primarily market-driven. While market dynamics drove the crisis as it occurred, the conditions that set the stage for the crisis to happen were to a significant extent formed by the state and coffee statecraft.

Assessing Vietnam's coffee "gamble"

It is no easy task to assess the successes and failures of Vietnamese coffee statecraft and its impact since the 1980s. As Peter Gowan has observed, economic statecraft is never a simple matter with a straightforward recipe for success, but often a risky "gamble."[7] There are many players involved, both domestically and internationally; the capitalist economy is intensely competitive and uneven; the outcomes are uncertain or unclear; and global market, political, and environmental events and patterns are highly unpredictable. It was within this context that Vietnam, burdened by the weight of decades of colonialism and the devastation wrought by the Vietnam War, had to find ways to construct its own comparative advantage against rich and powerful states, including new emerging Southern giants, like Brazil, China, and India. There was no guaranteed road to success and no assurances that a viable new export industry could be constructed in global markets dominated by established players – certainly, a great many poorer Southern countries have tried and failed to do so. It was from this position that Vietnam gambled on coffee as part of a broader, complex and contradictory, process of economic reforms.

One issue that is beyond dispute is that Vietnam's coffee statecraft successfully launched it up the coffee pyramid with historically unprecedented speed in a way that few had predicted, and in a manner that still surprises many today, who do not know that Vietnam accounts for nearly a quarter of all global coffee exports. Employing intensive, high-input production based on the widespread use of chemical fertilizers and irrigation, Vietnam came to have some of the highest coffee yields in the world – the average coffee yield in Brazil in 2013 was around 1,200 kg/ha, compared to 2,000 kg/ha in Vietnam.[8] This allowed it to surpass long-time coffee export leaders, inadvertently driving a global crisis. When the worst years of the crisis passed, Vietnam remained at the top of the volume chain, continuing to expand production and firmly established as the world's number one exporter of Robusta beans and number two exporter of coffee beans overall. In 2012, Vietnam exported well over three times as much coffee as Colombia and had moved astonishingly close to the export levels of Brazil, which has been the world's number one exporter for well over a century: in 2012, Brazil accounted for 25 percent and Vietnam 23 percent of the world's coffee exports.

Within Vietnam, the province of Dak Lak had been forever changed, with coffee fuelled growth facilitating a major increase in population, from 35,000 people in 1975 to over two million by 2003. On a national scale, coffee had become Vietnam's second most valuable export crop, after rice, covering 4.16 percent of the country's agricultural land, and providing employment for 3 percent of Vietnam's rural labor force. Vietnam's coffee farmers and workers combined number around 2.6 million today. To put this in perspective, it is over three and a half times the number of certified fair trade coffee farmers in the world. In short, Vietnamese coffee statecraft has succeeded in making Vietnam a world leader in coffee exports and in making coffee a major source of income, employment, and foreign exchange earnings for the country.

Importantly, Vietnam has not been alone in using statecraft to carefully manage the expansion of its coffee industry in recent decades. India, which has been involved in coffee exporting since the nineteenth century, has successfully employed the tools of statecraft to improve its position in the global industry. Even while carrying out relatively extensive neoliberal reforms on a national scale throughout the 1990s, which included removing the state-mandated monopoly held by the Coffee Board, the state continued to provide an array of extension services and market information to small farmers, while promoting coffee quality and consumption. The Indian state has been particularly devoted to research and development, and is considered "to be a world leader in coffee science."[9] As a result, India has successfully climbed the coffee ladder while others have fallen, rising from being the world's tenth largest coffee exporter in 1990 to the world's

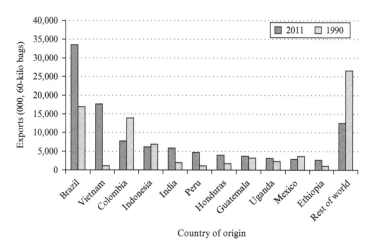

Source: ICO statistical database (http://www.ico.org), accessed July 30, 2013.
Figure 4.1 World's top coffee exporters, 1990 and 2011.

fifth by 2011 (see figure 4.1). As with Vietnam, India's position in the global coffee industry improved during the years
of the coffee crisis. This has been true in general for Asian
coffee exporters, who combined now constitute the world's
number two coffee region, exporting three and a half times
as much coffee as Africa. Vietnam alone exported over twice
as much coffee in 2012 as the combined total of all African
coffee exporting countries.

Beyond Vietnam's achievements in expanding its coffee
industry, assessing the developmental impact, and the human
and environmental costs and benefits, can be much trickier.
The growth of the coffee industry did help Vietnam diversify
its economy, an important factor in limiting its dependence
on a limited range of exports and promoting a more resilient
national economy. Vietnam's relative success in areas outside
of agriculture meant that by the 2000s the country's economic
growth was driven predominantly by industry and services,

which combined accounted for 78 percent of the country's GDP, compared to 22 percent for agriculture. The rural population, however, remained high and agriculture continued to be the largest employer, accounting for 65 percent of the workforce. Vietnamese farmers, moreover, have been comparatively poor and in possession of smaller plots of land in global terms; around 85 percent of Vietnam's farmers control land smaller than 2 hectares, and only 1 percent control plots above 5 hectares. Finding a substantial new avenue for agricultural exports with coffee has been of key importance for the Vietnamese state to slow the flow of rural-to-urban migration, to promote political stability in rural areas, to buffer the legitimacy of the state and state agencies, and to provide income for small farmers, rural workers, and the state.

Income and employment alone, of course, are not worth that much if they are insufficient to meet human needs and fight poverty. During the early boom years of Vietnamese coffee prior to the global crisis, the World Bank records generally improved living standards throughout Dak Lak, measured by growth in household incomes and decline in poverty rates. These initial gains were then significantly tempered by the coffee crisis, which devastated incomes and sparked unemployment in Vietnam as much as anywhere in the world. The crisis also weakened the capacities of the state. Burdened by declining revenues and increased debt – due in part to the state-mandated freeze on loan repayments and strategic tax deductions to prop up the industry – the state eliminated or reduced a number of health and education services, exacerbating social crises throughout the coffee regions. After a decade of steadily improving social indicators, the Central Highlands experienced no significant improvement from 1999 to 2003. The World Bank notes that this differed substantially from the rest of Vietnam, where the national poverty rate declined from 37 percent in 1998 to 29 percent in 2002. In the Central

Highlands, poverty remained high at 50 percent, with 30 percent experiencing hunger and malnutrition.[10]

Global coffee prices did eventually begin to recover, however, and along with them so did the Vietnamese coffee sector. The coffee indicator price climbed over $1 per pound in 2007 for the first time in nearly a decade. Global market forces then triggered a boom in 2011, leading coffee farmers around the world to plant new trees and step up the use of fertilizer and new technologies to boost production. Seeking to address its vulnerabilities to the coffee cycle, the Vietnamese state has taken measures to promote diversification within the industry, offering free land, cheap loans, and technical support to farmers to encourage them to grow higher-quality, and higher-priced, Arabica beans. So far, these initiatives have attained only modest success, with Arabica representing only around 3 percent of the country's beans. It is in Robusta production that Vietnam has constructed its comparative advantage, and it is with Robustas that Vietnam continues to find itself well situated. Robusta exports are growing more rapidly than Arabica each year, and Vietnam is exceptionally well positioned to gain advantage from one of the fastest growing Robusta coffee markets in the world, China – in 2011, Vietnam supplied 75 percent of all Chinese coffee imports. As a result, coffee continues to be a key export for Vietnam and a major rural income generator, with the average Vietnamese coffee farmer in 2012 earning more than the national per capita income of $1,300 per year.[11]

The relative income gains associated with Vietnamese coffee expansion are not without important shortcomings. First, as is generally the case with export-led growth in the commodity sector, the Vietnamese industry remains highly vulnerable to the ups and downs of the volatile coffee roller-coaster. By the end of 2013, global bean prices declined once again, in response to slowing economic growth in major coffee markets, with the composite indicator price dropping to its lowest value in several years. This raised the prospect of another crisis in the industry, leading for calls from within Vietnam for the state to stockpile as much as one-fifth of its production to prop up prices.[12] These fears were then temporarily allayed by another boom in prices in 2014 caused by a drought in Brazil, the long-term outcome of which remains uncertain. Second, the economic gains from coffee have been distributed unequally and often unjustly. During the initial phase of coffee expansion, indigenous groups in Dak Lak were forced off of their land, displaced by the state as part of the transition toward state and semi-private landholdings. Today, ethnic minorities continue to have lower incomes and higher poverty rates as a result of discrimination in granting access to quality land and agricultural supports.

Third, the ecological costs of Vietnam's coffee boom, driven by the widespread use of chemical fertilizers and irrigated water, mostly under full-sun conditions, have been high and it is not yet clear what the full environmental impact will be in the future. When coffee production soared during the 1990s, the use of chemical fertilizer more than doubled. This expansion occurred under generally lax oversight and regulation, raising serious concerns about long-term water and soil contamination. The use of water for irrigation also grew substantially so that by the end of the 1990s, 90 percent of the country's water use was for agriculture and 64 percent of agricultural land was under irrigation. This has given rise to growing concerns around water scarcity. Coffee production not only intensified, but also expanded, resulting in significant rates of deforestation; in the 1990s, forested areas in the Central Highlands declined by 19 percent. Much of this land was not optimal for coffee growing, which had the effect of further intensifying the use of chemical fertilizers and irrigation and, by extension, the extent of water and soil contamination, soil erosion, and water scarcity.[13]

Such impacts certainly point to the serious limitations of Vietnamese coffee statecraft and give pause to any overly optimistic appraisal of its developmental effects. There has, in fact, been little celebration of the rise of Vietnamese coffee among most international institutions, Western development organizations, and Western media.[14] In the aftermath of the global coffee crisis, Vietnam's role in flooding world markets with unprecedented quantities of cheap Robusta beans was often criticized for its shortsightedness and negative effects on other coffee countries. Given Vietnamese coffee's mixed impacts on reducing poverty and inequality domestically, and its negative ecological effects, most commentators have adopted a cautious or critical overall assessment, and not without good reason.

Despite its limitations, however, critical assessments of Vietnamese coffee statecraft are misguided if aimed primarily at policy makers and the state elite within Vietnam, as opposed to the wider structural imperatives of the global economy and the specific dynamics of the coffee industry. Real-world markets are not open fields of abundant opportunity, ripe with possibilities, easy and straightforward to access and compete in. Instead, they are intensely competitive at the international level, with trading, processing, and retailing dominated by a handful of oligarchic corporations, and production and supply dominated by long-established countries possessing decades of experience, and often significant political and economic weight. Seeking to break into established markets means, to a certain degree, succumbing to the rules of the dominant game, adapting policies and actions to meet the imperatives of global capitalism and the unpredictable vagaries of world market forces. These forces are beyond any single state's control, even while they exist in an international system managed by states that contend with each other in the interest of economic statecraft. The resulting competitive pressures lead to situations in which social crisis for some – such as the global coffee crisis – can result in economic advantage for others.

Vietnam was more or less compelled to take a "gamble" and find a means to fight its way into established global markets to meet ever-intensifying imperatives stemming from its territorial and capitalist logics. The expansion of a capitalist economy based on semi-private landholdings gave way to a class of export-dependent farmers and rural workers seeking income and employment; an increasingly urban, industrial economy seeking domestic markets to sell or lend to; and a state ever more dependent on its ability to find export commodities to gain foreign exchange, extract tariff and tax revenues, and promote political stability. Under these conditions, Vietnam's political elite rolled the dice on coffee, although the extent to which they did or did not understand the full impact of the gamble they were taking may never be clear. According to the World Bank's thorough investigation into the coffee crisis, "Vietnam may or may not have historically determined that the short-term pain of such an expansion might be worth the long-term gain in international market share."[15]

Conclusion

The assessment in this chapter reveals how the state played a key role in driving both the rapid emergence of Vietnamese coffee and the global coffee crisis. Contra the assumption of many free traders, market patterns in the world coffee industry are not primarily determined by the free flow of supply and demand, but are managed and to a large extent driven by economic statecraft. It is not just that states intervene in markets, but rather that they set the context in which markets function. The capitalist economy and international trade require states to enforce private property along with countless other laws and regulations. Once the rules are in place, of course, market forces are extremely powerful, unleashing often-overwhelming market imperatives that can be stronger than many individual states. The Vietnamese state may have played a central role in the expansion of its coffee industry, but once global market forces were unleashed and the crisis took root, it could not simply put the genie back in the bottle. Instead, it had to ride out the crisis like other coffee states.

The case for recognizing the centrality of the state should not be taken to suggest that the state is always "good" – in fact, the majority of states have very poor records in managing coffee statecraft in the interests of the majority of farmers and workers. And yet, this actuality does not remove the fact that the state is always *there*. Regardless of whether or not trade economists argue that there are efficiency gains to be made from the elimination of the state from the market, such an occurrence is impossible in the context of a capitalist economy, which requires states to ensure its very existence. This means that economic policy designed to combat poverty and injustice in the global coffee industry must take into account the centrality of the state if it is to have the possibility of substantive and long-term impacts.

The collapse of the ICA in 1989 did not mark the end of state involvement in the coffee market, but rather a transformation in relations between coffee states and the degree or intensity of competition between them. From a certain level of collective action during the 1960s–80s, coffee states shifted in the 1990s toward more competitive action, intensifying their efforts to gain comparative advantage over each other. Picking up steam throughout the 1980s, Vietnam proved to be particularly well situated and adept at taking advantage of the new conditions, bursting into coffee markets in the 1990s in a manner that sparked global chaos, only to end up as the world's indisputable second largest coffee exporter when the dust settled. Whereas the previous ICA quota regime would likely have provided collective checks and slowed the pace of Vietnam's entry, the post-ICA regime encouraged and intensified competition between individual states.

Reflecting on the political dynamics that have underpinned Vietnam's economic success, the long-standing advice by free trade think tanks and corporate lobby groups on how to address poverty and insecurity in the coffee industry (through individual or isolated efforts to improve productivity, quality, and diversification, as well as expand consumption) would seem of minor consequence when compared to the major impacts on world prices and trade patterns caused by the coffee statecraft of competing capitalist states vying for economic advantage. Such advice can be useful to individual states and to individual farmers – if they have the resources and capacity to act on it, which is generally the privilege of the wealthy few – as it can allow them to better compete against each other. But it does not resolve or address the wider political, social, and economic issues at play in the global coffee economy.

Progressive think tanks, nongovernmental organizations, and social justice advocates have, for their part, better understood the impact of the decline of the ICA on intensifying competition, price swings, and market chaos. But these groups have also too often accepted the notion of the triumph of the market over the state, overemphasizing the apparent decline of state power. As a result, they have increasingly turned their intention away from the state, perceived to be weak and ineffective, and toward efforts to influence corporate behavior directly by pressing for fair trade, organic, sustainable certification standards and better corporate social responsibility. This shift is, in many ways, understandable, especially with state officials perpetually claiming they lack the ability to control global markets. Moreover, in the era of transnational capital, corporate giants *do*, on the surface, dominate the global economy through their immense, and historically unprecedented, economic weight, global reach, and political influence.

This dominance, however, is ultimately exercised through complex layers of institutions, rules, and regulations put in place and managed by states domestically and internationally. Corporations acting on their own must meet the broader demands imposed on them by the global economy, the conditions of which continue to be steered by the territorial and capital logics of states. Nowhere is this more apparent than during the past two and a half decades of "free trade" in the coffee industry. During this time, state policy in Vietnam – a lower-middle-income country with an economy ranked 55th in the world – made a major impact on global market conditions and the livelihoods of millions of coffee families.[16] The global shift from collective action to intensified competition among coffee states that has characterized the free trade era in the coffee world has been accompanied with a general denial among state officials of their ability to control markets, even while their actions have continued to say otherwise. Under these political conditions, social justice advocates have increasingly felt the need to turn away from states and toward market-driven, nongovernmental certification schemes and corporate social responsibility to address poverty and inequality in the coffee industry. The results, as chapter 5 describes, have been important gains for the ethical trade movement, but of only modest reach and breadth, and with little overall impact on the structural roots of coffee's deep, uneven patterns of trade and production.

CHAPTER FIVE

Fair trade and corporate power

Given commonly held views of the modern state as unresponsive, inefficient, undemocratic, or ineffective at substantially
addressing concerns around inequality, poverty, and underdevelopment in the coffee industry, social justice groups,
nongovernmental organizations, international institutions,
and government itself have turned increasingly toward the
market as a tool for guiding progressive social change through
various forms of "ethical consumerism" – fair trade, organic,
sustainable coffees and an array of corporate social responsibility programs. Through market action, consumers are seen
as being able to influence corporate behavior in a manner
beneficial to coffee farmers and workers, while bypassing the
inefficient, repressive, or disinterested state.

The previous chapters already raise some warning flags for
the general understanding of the global coffee market implicit
in this view. First, the rise of ethical consumerism must be
understood not merely as a strategic attempt to find new
avenues to promote development in the coffee sector, but as
the flip side of the expansion of neoliberal reforms since the
1970s and the gradual decline in the commitment of states to
socially regulate the coffee market domestically and internationally. The rise of ethical consumerism is part and parcel of
the rise of neoliberal governance, wherein legally mandated,
state-directed forms of social protection are increasingly abandoned in favor of private, market-driven, voluntary initiatives.

Second, the notion that states are being bypassed through

direct engagement with corporations is, in many ways, misleading, as corporate power is intricately interwoven with
the capitalist state and vice versa. Through a territorial logic,
states work to defend and promote the interests of nation-
ally based corporations abroad, perceived as essential for the
state in providing revenue, employment, and economic and
social stability, while corporations in turn exercise a dominant
influence in determining state priorities abroad through their
overall economic weight, control of major media and information outlets, and direct influence over government through
political contributions, donations, and bribes. Through a
capitalist logic, the very rules and regulations created and
enforced by capitalist states give rise to the existence of corporate entities, which in turn work to further modify state
policies in the interest of private profits. Since the early 1980s,
these two logics, in the context of declining profitable investment opportunities for Northern-based companies, have
compelled powerful capitalist states to more aggressively seek
out new markets, investment opportunities, cheap labor, and
resources abroad, while cutting social spending and corporate
taxes, and privatizing previously underexploited or commonly
held resources.[1]

Within this context, critics have argued that the ethical
consumerist movement has emerged as a form of limited,
market-driven, ad hoc social protection on the cheap to
replace the existence of and aspirations for broader, more universal, more extensive social protections typically associated
with state-managed activities. This differs sharply from the
more "empowering" vision of ethical consumerism offered
by its diverse promoters. Among advocates, one often hears
reference to the existence of "consumer society" or "consumer culture" or "consumer sovereignty," wherein industry
is driven by consumers who ultimately use their "consumer
power" to determine the market patterns that private compa-

nies merely respond to. From this perspective, the existence or lack of socially and ecologically just methods of production and trade in coffee and other industries is not primarily a result of decisions made by corporations and states, but an outcome of consumer behavior – collectively, consumers are "sovereigns" who drive corporate action through their purchasing decisions.

Far from the consumer being king of the market, however, consumers are generally called upon to exercise their consumer power without access to meaningful or substantive information on where a product is produced, under what conditions, and with what social and ecological impacts. Instead, they make market determinations under the manipulation of massive corporate advertising campaigns designed to engineer consumer choices before individuals ever hit the market in the first place. By one calculation, in the United States alone big business spends over a trillion dollars a year on corporate marketing – over 20 times the annual amount estimated by the UN required to eliminate world hunger.[2] Under these conditions, middle-class consumers on a global scale crowd into supermarket chains, saturated with corporate images and messages, to select from an artificial array of branded goods produced in similar ways by an increasingly narrow range of corporate giants – what sociologist Anthony Winson refers to as "pseudo-variety."[3]

Coffee consumption would appear to take place not within the context of a much celebrated "consumer culture," but rather within a corporate culture. After a long and continuing process of mergers, bankruptcies, and acquisitions, the global coffee chain has become dominated by a handful of corporate juggernauts, with just five roasters responsible for purchasing nearly half of the world's supply of green coffee beans: Kraft Foods Group, Nestlé, Sara Lee, Procter & Gamble, and Tchibo. These companies' oligopolistic control over access to core Northern consumer markets has given them the power to manipulate prices on the global coffee market, and their immense size has given them significant advantages over domestic competitors due to their economies of scale, access to new technologies and innovations, and massive marketing budgets and "brand power." Increasingly since the 1970s, giant roasters and retailers have pushed the industrialization, mass distribution, and homogenous consumption of international brands through multimillion-dollar marketing strategies.

And yet, it is not only corporations that have vied for consumer loyalties. Growing up alongside the further expansion of corporate power has been the specialty coffee industry, which has offered niche markets for those seeking better ethical, environmental, and health standards, including various forms of fair trade, organic, and sustainable coffees. The best of these initiatives, of which fair trade certification should be considered the industry ethics leader, have been built through substantial "bottom-up" involvement from social justice groups working with small farmer organizations and have attained limited but important social gains for the participants. These gains, as I will argue, have not been able to match the breadth and impact of state-driven projects, and there is evidence in recent years of fair trade being swallowed up or shuttled aside by watered-down, big money, corporate social responsibility initiatives. While ethical consumerism has grown significantly in popularity in recent years, it is not clear that this popularity has translated into the sort of substantive gains needed by the world's poorest coffee families, as fair traders must limit their goals to the confines set by a global market beyond their ability to control through "consumer power."

Coffee branding

Corporate power and wealth within the global coffee chain are based not solely on market share and oligopolistic dominance of roasting and distributing, however essential they are, but also on the ability to define the coffee identity, norms, and quality standards. Sociologists Benoit Daviron and Stefano Ponte have argued that market power in the coffee industry comes not just from struggling to gain the material value of the coffee beans, but increasingly from their "in-person" service and symbolic (subjective elements rooted in a commodity's reputation) qualities as well.[4] Developing and capturing these in-person and symbolic qualities has sparked ever more intensive and expensive marketing battles, ones in which coffee industry leaders have proven particularly skillful and successful at.

Central to the coffee industry's marketing success has been the ability of major roasters and retailers to tie coffee brands to deeply held loyalties and identities around gender, class, and nationalism. For example, throughout the twentieth century, gendered messages centered on a "good wife's" ability to serve up a "good cup" of coffee played an important role in coffee marketing campaigns. Journalist Mark Pendergrast describes one particularly revealing cartoon strip advertisement for Chase & Sanborn coffee in 1934. Rooted in gendered norms, and the acceptable terrain of domestic violence at the time, the ad makes clear the expectations of a good housewife, and the punishment or praise she might face depending on her success or failure in delivering a good cup of coffee to her husband:

> "Here's your coffee, dear," a wife says to her scowling businessman husband over the breakfast table. "I thought we were too old to play mud pies," he growls. Flinging the hot coffee at her, he yells, "What did you put in it this time?

Bricks or gunpowder? See how you like it!" She cries, "Oh, you brute! I'm all black and blue." In the final two frames she wears a catcher's mask and holds a shield while offering him a cup of Chase & Sanborn . . . . Of course the husband loves it. "Take off the mask, darling . . . . This is too fresh and good to waste a drop."[5]

Class identities have also long played a central role in coffee marketing campaigns, with large-scale fast food chains often targeting blue-collar workers and suburbanites, while specialty coffee roasters, like Starbucks, have emphasized white-collar, urban identities in their advertising and branding. For blue-collar workers, chain stores like McDonalds, Dunkin' Donuts, and Tim Hortons have carefully crafted their images as places where people from all social classes can meet and take part in small luxuries, like drinking a good cup of coffee, offering an affordable form of social leveling. For suburbanites, who frequently come from wealthy or middle-class backgrounds, the appeal of chain stores does not stem from social leveling. Instead, suburbanites – as well as blue-collar workers – have often been drawn to chain stores on the basis of what historian Steve Penfold describes as "a sense of ironic pride in the lack of cultural alternatives" in comparison to the perceived snobby, artsy, or hippie culture of big urban centers.[6] The most successful coffee and food chains have astutely recognized and incorporated these class expressions into their marketing campaigns, often interwoven with gendered symbols and, in some cases, intensely held shared values and myths of community associated with nationalism.

Perhaps one of the most telling examples of the power of nationalism in advertising is the coffee and donut chain Tim Hortons, the largest retail chain in Canada, owning over 3,300 restaurants in the country and accounting for 8 out of every 10 cups of coffee sold there. Named after its founder, Tim Horton – a deceased professional hockey player, frequently depicted as a physical player, but a "gentleman" – the giant chain has emerged as one of the country's most widely recognized symbols of Canadian nationalism. Through multimillion-dollar marketing campaigns first launched in the late 1990s, Tim Hortons has successfully rolled together manly, folksy, nostalgic notions around hockey in Canada and geared them toward consumption and corporate profits, appealing to Canadian nationalism and "tradition" in a country that generally has been overwhelmed by US culture industries and has historically experienced a weak popular nationalism.[7] Tim Hortons' marketing campaigns have been enormously successful, ensuring it steady profits and store expansion, and in many ways obscuring the fairly conventional, giant chain store nature of its everyday operations.

Beyond branding and advertising, it is important to keep in mind that most of the lead companies in the coffee industry have enhanced corporate profitability through the use of relatively low-paid, non-unionized workers to process, roast, package, distribute, and sell coffee. This has been the case in both the North and the South. In the North, coffee companies have drawn on low-paid, seasonal, or piecework employment to keep costs down in the roasting, processing, distribution, and marketing stages. In the South, Northern-based transnational companies have frequently owned and operated processing facilities for final cleaning, de-shelling, and classifying of green beans before export, also drawing on low-paid, highly exploited laborers. In both instances, women have often been subjected to the worst labor conditions. Due to highly gendered norms around male and female work, women's wages have often been deemed as "subsidiary" to male "breadwinner" wages, allowing women to be paid less and receive fewer benefits, making them desirable workers for the coffee industry. At the beginning of the twentieth century, women coffee workers employed at a Hills Brothers factory in the United States were paid less than half of their male counterparts' rates, worked 10 hours a day for six days a week, and were given only one week of vacation per year.[8]

While a great deal more can be said about workers in coffee processing and retailing industries, historically and today, two things are particularly important to keep in mind. First, although general labor conditions have improved over the past century, especially for Northern workers, millions of workers continue to confront poor working conditions, job uncertainty, and poverty. This is especially the case in the retail industry, which draws most of its workforce from "flexible" labor markets, characterized by low-paying, precarious jobs, limited health and pension benefits, a high annual turnover rate, and a comparatively high risk of unemployment. These jobs tend to be disproportionately concentrated in youth and "high-risk" groups (single mothers, new immigrants, racial and ethnic minorities, persons with disabilities, and adults with limited education), whose labor forms the backbone of the coffee industry in the North.[9]

Second, it is always important when assessing labor market trends to keep politics and power in mind. Coffee companies have not merely responded to labor market conditions, taking advantage of cheap labor when they can, but have often worked actively and aggressively to attain these conditions, lobbying governments, funding politicians and political parties, controlling the message on labor rights through their influence over advertising and the corporate media. Corporate interests are deeply interwoven with those of the capitalist state and giant coffee companies play important roles in pressing governments to develop policies that lower real minimum wages, employment and welfare benefits, corporate taxes, and labor and environmental protections, while aggressively fighting against unionization and workers' rights to collective bargaining within their firms. This is the case even for those coffee companies with the best record of "corporate social responsibility (CSR)." Chief among them is Starbucks Coffee Company, the world's largest specialty coffee roaster, as well as one of the world's largest fast food chains, operating over 19,000 stores in 62 countries, employing over 200,000 people, and making $14.9 billion in revenue in 2013.[10]

Since its creation and rapid growth beginning in the 1970s, Starbucks has emerged not only as a major global coffee roasting chain, but also as a leader in CSR, winning countless awards and praise, predominantly from the business sector and corporate magazines. As neoliberal reforms have gradually eroded the commitment by states to socially regulate the actions of TNCs at home and abroad, the TNCs have stepped up their efforts around CSR, seeking to demonstrate that a more self-regulatory approach to corporate governance can address social and environmental concerns, as well as to counter specific accusations against them by global justice groups and protect their brand images. For Starbucks, this includes an array of international "sustainability" projects (including fair trade certification) as well as a payments and benefits plan for full- and part-time retail employees (whom Starbucks calls "partners"). Benefits to retail workers include higher wages than among most fast food competitors; basic medical, dental, and vision coverage; basic mental health and dependency treatment; and a variety of small additions, such as free beverages at work or a free pound of coffee per week. Consequently, Starbucks employees are better off than workers in competing retail or coffee chains, which generally have much higher employee turnover rates; Starbucks' annual employee turnover rate has been estimated at around 60 percent, compared to the industry average of an extraordinarily high 150–200 percent.[11]

And yet, while most workers would understandably prefer a job at Starbucks to poorer "McJobs" in other fast food chains, Starbucks jobs still remain relatively low-paying and precarious compared to most other sectors of the economy (its employee turnover rate still means that less than half of Starbucks workers last one year in employment), and Starbucks' actions outside its CSR program have often been directed against broader and more substantive labor rights and social welfare provision at the state level. Despite offering top-down concessions to its workforce, Starbucks is fiercely anti-union. In one particularly telling example, Starbucks aggressively fought off unionization efforts at 10 Starbucks stores in Vancouver, Canada, from 2000 to 2005. Originally, workers at the store successfully unionized under one collective agreement as members of the Canadian Auto Workers, Local 3000 (CAW 3000). Workers negotiated two collective agreements with Starbucks, both of which provided minor concessions to workers regarding issues of seniority, wage improvements, and scheduling. Fearing the precedent this was setting, Starbucks refused to negotiate a third agreement. CAW 3000 was placed in a legal strike position and organized an "UnStrike" campaign, encouraging consumers to shop only at unionized Starbucks stores in British Columbia. Given the vulnerability of the workforce and the high employee turnover rate, Starbucks merely refused to negotiate and waited the campaign out until its main organizers left and the union collapsed.

Beyond its anti-union approach to labor relations, Starbucks has also participated in an array of corporate activities that call into question its broader commitment to social responsibility. In 2012, Starbucks UK came under fire for being a major player of corporate "tax avoidance" – a process whereby corporations use a variety of complex loopholes in the tax system to pay little or no tax. Despite making over £3 *billion* in sales in the UK in 14 years from 1998 to 2012, it was revealed that Starbucks paid only £8.6 million in corporate taxes – less than 1 percent of its sales figures. Public protest in the UK eventually compelled Starbucks to voluntarily agree to pay an additional £20 million in taxes. Such tax avoidance measures have the effect of denying the state significant public revenue, increasing the tax burden on middle-class income earners while starving public services of much-needed funds. An interesting and telling comparison to these tax avoidance measures are the political activities of Starbucks' billionaire CEO, Howard Schultz, who has spent considerable money and effort over the past several years publicly promoting government austerity and national debt reduction in the US – Starbucks management thus avoids paying corporate taxes on the one hand, while scolding government for spending beyond its means with the other.[12]

The necessity of putting shareholder value and corporate profits first means that Starbucks' CSR can only entail modest, top-down initiatives entirely within the corporation's control and oversight, while Starbucks continues to oppose wider "socially responsible" activities such as worker unionization and collective bargaining, and corporate taxation. Starbucks is certainly not unique in the coffee industry and, in fact, compared to other coffee giants is indeed deserving of the title of "ethical" leader. The problem is not Starbucks, but CSR itself. Regardless of what ethical intentions individuals managing the giant coffee transnationals may or may not have, in the end they run immense, hierarchical institutions designed with one primary objective: to enhance corporate profitability in order to survive, thrive, beat out competitors, and enrich private shareholders.

While the development and popularity of CSR are something specific to the current era of neoliberalism, its fundamental objectives remain rooted in the long history of corporate activity in the coffee industry. First, CSR initiatives are designed in a top-down manner to offer a minimal form of social protection to stave off criticism and counter bottom-up demands for worker unionization, collective bargaining, and more universal and extensive state-managed social, labor, and environmental regulations – all things that major coffee companies have long fought against while profiting from a relatively precarious workforce. Second, one of the primary objectives of CSR is to protect or enhance corporate brand images, which the major coffee companies have constructed over decades through persistent multimillion-dollar marketing campaigns. In the case of Starbucks, one of the keys to its marketing success has been to provide a comfortable, trendy café environment and a sense of local community in its stores. This requires attention not just to interior design and customer service, but also to public relations aimed at making consumers feel *good* about drinking Starbucks' coffee – as if they have connected with, rather than exploited, workers and farmers. Starbucks has been particularly adept at using CSR in this regard, actively promoting its "ethical" image in store displays, pamphlets, products, and online. CSR thus represents a new twist on two long-standing themes of the coffee industry: minimalist labor standards, and maximalist marketing campaigns. It is in these areas, by design, where CSR has its greatest impact.

The fair trade alternative

While CSR offers a top-down, shareholder-driven approach to ethical trade, it is not the only approach, and others have emerged and developed over the past decades that are more bottom-up and "stakeholder" driven, involving direct participation from the poorest and most vulnerable in the coffee chain. Beginning in the 1970s, an ever widening array of issue-based products have hit supermarket shelves, part of what sociologist Marie-Christine Renard refers to as "interstices in the midst of prevailing tendencies" in a global agro-food system characterized by industrialization, corporate monopoly, mass distribution, and the homogenous consumption of international brands. Against these tendencies, "niches and micro markets" have emerged in the North to appease a minority of consumers who distrust TNCs and desire products that express cultural diversity and "old-fashioned" nostalgia, while making a variety of health, "natural," organic, ecological, and ethical claims – with wide-ranging degrees of precision and accuracy.[13]

Among this growing ethical consumerist movement, fair trade certification has been a world leader, having among the greatest transparency and most rigorous standards. The fair trade network is a formal system of nongovernmental organizations that connects peasants, workers, and craftspeople in the South with partners in the North through a system of "fair trade" rules and principles. The network was first developed in the 1940s and 1950s on the initiative of alternative trade organizations that sought to provide assistance to poor Southern producers by creating a trading system in which prices would be determined on the basis of social justice, as opposed to unpredictable international markets. By many of the founding fair trade organizations, such as Oxfam International, the network was considered part of a broader movement that promoted a new international economic order based on strong state intervention at the national and international level to support development efforts in the South.

In the 1980s, the orientation of the network changed as fair traders moved away from the vision of an alternative trading system and instead sought to gain access to conventional markets. This shift was led by the emergence of fair trade labeling initiatives, coordinated under the umbrella organization Fairtrade International (FLO), which have sought to certify conventional businesses willing to meet FLO's fair trade criteria. According to FLO standards, fair trade goods are produced in the South under the principles of "democratic organization" (cooperatives for some commodities, like coffee, and freedom of association and collective bargaining for workers for other commodities where substantial numbers of small farmers do not exist, such as tea), no exploitation of child labor, and environmental sustainability. Goods are exchanged under the terms of a minimum guaranteed price, with social premiums paid to producer communities to build social and economic infrastructure.

The reorientation of the network initiated by fair trade labeling was, in part, driven by the desire to expand fair trade markets beyond their relatively small size. Equally important, however, were the changing political and ideological conditions signaled by the rise of neoliberal reforms and the decline of social regulation at the national and international level, including the collapse of the ICA in 1989. Seeing the writing on the wall, fair traders adopted a new, market-driven vision of fair trade based on non-binding, voluntarist commitments from private corporations. The result has been a significant success for the fair trade network, in terms of sales growth, which has increased substantially since the 1990s. This growth has been driven by the increasing participation of national and international bodies, such as the World Bank, as well as major TNCs, which view the fair trade network as a voluntarist alternative to more robust state regulation.

Although the fair trade network was traditionally based on handicraft sales, since the 1970s and 1980s coffee has been one of its most popular "flagship" products, riding a tide of growing popularity in Europe and North America for higher-quality, whole Arabica beans roasted by specialty coffee companies. Fair trade coffee's growing success has had a generally positive impact on hundreds of thousands of coffee farmer families who have been able to participate in fair trade certification. Overall, and to significantly varying degrees, the existing and growing body of research on fair trade suggests that Southern partners have often been able to attain better access to social services through cooperative projects in health care, education, and training, as well as enhanced access to credit, technology, and economic infrastructure (such as processing and transportation facilities) as a result of fair trade.

At the same time, one must be careful not to exaggerate the benefits of fair trade, whose Southern participants are generally better off, but remain relatively poor. Widely cited research on fair trade coffee in Southern Mexico by Daniel Jaffee, for example, demonstrates that while fair trade certified farmers have attained higher household incomes, the majority of additional income has been spent hiring extra labor to meet fair trade and organic standards. The additional income, argues Jaffee, has been diverted from the participating fair trade family, even while it still benefits the community overall in the form of additional rural wages. Similarly, extensive research conducted by Christopher Bacon in Nicaragua, involving interviews with 228 coffee farmers, has determined that fair trade farmers have experienced higher incomes and reduced "livelihood vulnerabilities." At the same time, prices paid to farmers are often still below the cost of production, over 70 percent of fair trade farmers reported a decline in their quality of life over the past few years, and cooperatives regularly have to sell up to 60 percent of their coffee on conventional markets due to the limited size of fair trade and organic niche markets. These sorts of outcomes are common throughout the fair trade network.[14]

There are, in fact, major limitations to the overall impact of fair trade coffee, and critics have pointed to numerous shortcomings. Perhaps most significant are the overall breadth and reach of fair trade, which are strictly limited by its dependence on relatively small niche markets in the North. While fair trade coffee has grown considerably since the early 1990s and now reaches over 670,000 coffee farmer families, this represents only 3 percent of the world's 25 million coffee farmer families. The same limits exist for the fair trade price, which is not negotiated between representatives of consumer and producer countries as the ICA was, but rather is based on a guarantee that fair trade prices will remain slightly above conventional prices with an additional assured floor price. (Since 2011, the fair trade minimum price for washed Arabica coffee beans has been 5 cents above the conventional market price with a guaranteed floor price of $1.40 per pound, plus an additional 20-cent social premium and an additional 30 cents for certified organic.) Thus, while fair trade prices are guaranteed to be above conventional ones, they remain tied to a range whose parameters are set by conventional prices and are not high in historical terms. From 1976 to 1989, the regular price of conventional coffee beans under the ICA system was close to, and in some years a fair bit higher than, what is considered the "fair trade" price (see figure 5.1).[15] Moreover, unlike fair trade, which reaches only a small minority of farmers, ICA-era prices reached all of the world's coffee farmers.

A second major consideration when assessing fair trade is the extra costs associated with meeting the many fair trade standards, which have primarily been developed and managed by Northern organizations on the basis of Northern norms, and are increasingly interwoven with an array of other standards set by wealthy states and transnational companies – such as organic certification or the European-based GLOBAL Good Agricultural Practices (GLOBALG.A.P.). While these standards are generally desirable, they are also onerous and costly, imposing significant extra burdens on small farmers who often have little or no say in their development. The fair trade network has, over the years, responded to this criticism by expanding producer representation on various governing

Source: UNCTAD statistical database (http://unctadstat.unctad.org), accessed July 30, 2013.

Note: The fair trade minimum price is the FLO price. If conventional prices go above the minimum price, the fair trade price moves 5 cents above the conventional price plus the social premium.

Figure 5.1 Comparing fair trade and conventional coffee prices, 1960–2012.

boards, but overarching concerns around producer input have remained and cannot be easily resolved. To a large extent, this is due to the very nature of market certification, which generally involves a pre-given set of standards and an outside verification body to provide reliability and transparency to ethical consumers. Producer groups, however, are often left out in the cold, which, as argued by sociologist and union activist Henry Frundt, can be seen as an affront to democratic labor principles around "freedom of association," which at its core requires that labor conditions are negotiated and audited by producers themselves – not by an external party over which people on the ground have very little influence or control.[16]

Third, in broader developmental terms, critics have

expressed concern that fair trade promotes continued dependence on tropical commodities and the vulnerability and
uneven development typically associated with the regular
operations of the global market. In response, defenders have
correctly argued that most small producers do not have viable
alternatives to tropical commodity production and that those
that do still require the support of fair trade standards to assist
them in their transition to other economic activities. At the
same time, the overarching concern here is also true: fair trade
is inherently limited by its market-driven structure and its
dependence on the current, highly unequal patterns of global
trade and consumption. As is the case with conventional consumerism, fair trade consumerism remains premised on the
notion of "consumer sovereignty," discussed in the opening
paragraphs of this chapter. While fair trade does challenge
the assumption that consumers have adequate information
upon which to base their market choices, it remains rooted
in the belief that these same consumers should have the final
say in how goods are produced and distributed. The effect is
to reproduce a world trading system in which the needs of
relatively poor Southern producers remain subservient to the
demands of Northern consumers.

Relying entirely on the market to mediate relations between
producers and consumers in a highly uneven world significantly waters down the democratic, participatory, and
redistributionist aspects of fair trade. Within the North, the
consumer base upon which fair trade relies is composed of
relatively well-off consumers whose income is ultimately
derived from an unequal distribution globally and nation-
ally. Ecologically, fair trade depends on existing consumption
patterns characterized by "overconsumption" in the North
that contribute disproportionately to climate change and
global environmental stress. Perhaps most significantly, the
ethical consumers that drive fair trade remain isolated indi-

viduals whose primary responsibility is to shop. Fair trade's democratic principles stop and end with Southern producer cooperatives, leaving global linkages to be tossed and turned in the global market. Consumers are not connected with producers through a democratic or genuinely participatory process and possess only "purchasing power" to influence fair trade and global markets generally. Their knowledge of the lives of producers is limited to fair trade advertisements, mediated by the market in a manner that shields them from direct contact and from direct responsibility for the outcome of consumption choices. In this context, individual preferences drive ethical consumption, not a *shared responsibility* for purchasing decisions and their social and ecological impacts. This places significant limitations on the price and reach of fair trade and leaves human development for hundreds of thousands of coffee farmers dependent on an "ethical premium" that consumers – not themselves directly impacted by good or bad purchasing decisions in their daily lives – may or may not be willing to pay from one day to the next.

As a final consideration, critics have raised growing concerns that the growth of fair trade since the early 1990s has been driven increasingly by powerful institutions that use mild support for fair trade to obscure their wider devotion to "free trade." The World Bank, for example, has given increasing support to fair trade – in policy documents, but also by serving fair trade tea and coffee to its employees in Washington, DC – while continuing to push ahead in international affairs with a free trade agenda that has generally had destructive impacts on small farmers and rural workers. Similarly, Starbucks attains positive public relations for selling 8.1 percent of its coffee beans fair trade certified, even while over 91 percent of its beans are *not* fair trade certified and the majority of its Northern workers are non-unionized and less job-secure than non-retail sectors. This contrasts with the traditional fair trade organizations that originally built the network. In Canada, for example, three original fair trade leaders, Planet Bean Coffee, La Siembra, and JustUs! Coffee, are all worker-owned cooperatives devoted to selling 100 percent fair trade certified products, generally pay farmers above the fair trade minimum, and are often dedicated to promoting fairer labor rights in both the South and the North and to educating consumers about the inequalities in global trade.[17] Alternative trade organizations such as these, while they continue to do well on their own terms, are being increasingly crowded out by conventional corporations that are emerging as the dominant players in fair trade and other forms of ethical consumerism.

Corporate social responsibility and fair trade: who is changing whom?

Fair trade coffee certification over the past few decades has proven to have a meaningful but at the same time limited impact on global coffee markets. It is limited by its relative inability to directly influence the decisions of states that manage the global coffee economy – while many fair trade groups have organized themselves to put pressure on national governments to respond to their needs, the fair trade system overall is not devoted to this end and is increasingly evoked by powerful institutions as evidence that voluntary, consumer-driven action can replace state social regulation. No doubt many fair traders would agree with this critical assessment, seeing the turn toward ethical consumerism as a strategic necessity in a world where states have continued to carry out neoliberal reforms and abandon social reform.

Perhaps more surprising than fair trade's limited impact on states, however, has been its still highly limited impact on corporate behavior. Since the emergence of fair trade labeling in the 1980s, one of the key goals of the network has been to change corporate behavior by demonstrating that real ethical markets exist and that, should corporations want to maintain or gain access to them, they had better get on board with fair trade certification. To some degree, this has occurred, as seen by the blossoming of countless new ethical and sustainable trade initiatives, often developed in response to the fair trade network, which has raised the ethical bar for the entire industry. At the same time, much of the new corporate "buy in" involves ethical initiatives that are either weaker versions of fair trade or have the potential to water down fair trade certification itself.

Starbucks perhaps best represents this complicated scenario, having adopted fair trade while at the same time limiting fair trade's broader vision and standards. While purchasing 8.1 percent of its beans certified as fair trade, Starbucks has devoted significantly more attention to its own Coffee and Farmer Equity (CAFE) program, through which it has essentially self-certified the vast majority of its own coffee beans – playing roles as both a key participant in fair trade and its major competitor. While not without its positive aspects, Starbucks' CAFE is essentially a weaker version of fair trade certification: whereas fair trade involves genuine independent third party certifications, Starbucks has developed its own standards, in consultation with Conservation International, and hand-picks and trains its own verification agents; whereas fair trade involves a guaranteed minimum price, CAFE offers a less clear commitment to a higher price and preferred access to sell to Starbucks; whereas fair trade coffee certifies only small farmers, Starbucks' CAFE includes giant coffee plantations that already possess significant advantages over small farms; and whereas fair trade requirements include minimum standards that *all* certified partners must meet, Starbucks' CAFE offers a grading scale based on how well a partner scores in meeting basic standards. The last of these is highly significant, as it means that CAFE includes as "certified" any farm that meets only basic initial requirements. Thus, while Starbucks claims that 84 percent of its farmers in 2010 were CAFE certified, nearly 60 percent of them had attained only "verified" status, meaning they had scored *less than 60 percent* on their CAFE evaluations. One could thus ignore any number of standards – collective bargaining, freedom of association, worker safety and training, access to education and medical care – and still be considered a "verified" participant of the CAFE program.[18]

In addition to its role as a fair trade competitor, Starbucks has also impacted fair trade standards from within. While purchasing only a small portion of its beans fair trade certified, this amount has made Starbucks the largest single seller of fair trade in North America, giving it significant influence over the network's future direction. The impact of this influence became extremely apparent in 2011 when TransFair USA renamed itself "Fair Trade USA" and split with FLO, widely regarded as one of the world's most rigorous and legitimate international fair trade umbrella organizations. The biggest issue of contention between the organizations was the role of giant coffee plantations in fair trade, which FLO does not certify. Although it is not clear how much of a role Starbucks played in the actual split, Starbucks and other corporations have long pushed for this rule to be eliminated so that they can expand certification with traditional established partners, giant coffee plantations. FLO has resisted these demands for years, insisting that there are millions of small-scale coffee farmers who desperately need the support of fair trade. In breaking with FLO, Fair Trade USA declared its keen readiness to certify coffee plantations. Whether explicitly or implicitly, Starbucks' influence played a major role on these events and on the gradual watering down of fair trade standards that they signify.

Finally, Starbucks' impact on fair trade extends also to the wider political vision traditionally associated with fair trade. The fair trade network was originally constructed by a diverse array of social justice and development organizations that viewed fair trade consumerism as part of a wider political project to raise awareness around the injustices of global trade and promote an alternative trading system. Starbucks, in contrast, has adopted fair trade as only one component of a wider CSR agenda that is far removed from the moral and political mission of fair trade and in many ways obscures the injustices in the global trading system. Despite promoting fair trade and a variety of market-driven sustainable coffee initiatives, the necessity of putting shareholder value and corporate profits first has often pitted Starbucks against the broader needs of poor coffee farmers and workers.

One of the most notable examples of this took place from 2005 to 2007, when the Ethiopian government took measures to trademark its high-quality coffee beans to bring a greater share of coffee value to Ethiopian coffee farmers. This effort was met with resistance from the US National Coffee Association (NCA) and other industry lobbyists, which put pressure on the US Patent and Trademark Office to reject or delay approval of the patent. Some industry critics suggested Starbucks played a role behind closed doors in pressuring the NCA to adopt such a tough stance, given its reliance on the sort of high-quality beans grown in Ethiopia, a claim that Starbucks denied. Starbucks did initially reject the idea and publicly renounced it through a media campaign. In response, Oxfam International launched its own campaign in support of the initiative, persuading over 96,000 people to contact Starbucks through e-mails, faxes, phone calls, postcards, and in-store visits. Feeling a mounting threat to its brand image, Starbucks gave in and in 2007 signed an agreement that it would use and promote Ethiopian coffee beans in accordance with agreed terms and conditions of the trademarking initiative (discussed further in chapter 6). After that, Ethiopia successfully registered its trademarks in the United States and around 30 other countries. While the full extent of Starbucks' resistance to the initiative may never be clear – whether it involved their publicly stated reluctance or additional efforts through corporate lobbying – the overall scenario may give us pause in considering Starbucks' claims of global social responsibility: the patent, at the most, can offer Ethiopian farmers only an additional few cents of the final sale price of a bag of coffee, and Ethiopian farmers are among the poorest in the world – Howard Schultz's total compensation as CEO of Starbucks was $29.73 million in 2011, equivalent to the annual income of over 83,000 Ethiopians.[19]

Conclusion

In the end, Starbucks' commitment to fair trade and "social responsibility" entails relatively modest, top-down initiatives entirely within the control and oversight of the corporation itself, while it continues to oppose wider socially responsible activities, whether they involve coffee farmers in the South or unionization efforts among Northern employees. In this, Starbucks is not unique and, in fact, compared to other coffee giants, it is deserving of the title of leader in CSR. The problem is CSR itself, which is not designed to meet the needs of a broad public for universal social programs and policies, but rather to offer the most minimal form of social protection possible to stave off criticism and protect or enhance corporate brand images. This should raise major cautionary flags for fair traders given the growing convergence between CSR and fair trade. The vast majority of Northern consumers do not know the difference between fair trade and other ethical consumer projects, corporate-driven or not, giving corporations considerable flexibility in their CSR strategies and allowing them to offer modest or token support to fair trade while developing their own pro-corporate initiatives and putting pressure on fair traders to weaken their standards. As more and more corporations buy in to fair trade coffee – now being sold by an array of transnational companies, including Wal-Mart, McDonalds, Dunkin' Donuts, and Nestlé – fair trade's wider political vision gets lost and replaced with a narrower form of corporate giving within the confines of the existing global coffee industry. This pulls fair trade further into corporate dependency, to subsuming the demands of poor workers and farmers to the profits of wealthy shareholders, and to accepting the dominant consensus around the "market" as the only solution to poverty and inequality, with the "state" receding ever further into the background. The effect is not merely to overlook the power of states to confront long historical patterns of poverty and inequality in the coffee industry, but to let states off the hook for their central role in creating and reproducing these patterns and allowing them to persist in new and varied forms.

CHAPTER SIX

Coffee and the non-developmental state

In 1993, just four years after the end of the ICA, over a dozen coffee exporting countries, accounting for 70 percent of global supply, formed the Association of Coffee-Producing Countries (ACPC): a cartel designed to regulate supply and halt rapidly declining coffee bean prices. Three of these countries – El Salvador, Costa Rica, and India – had previously opposed the ICA, thinking its collapse would lead to better, not worse, coffee incomes for their farmers. After some success in propping up prices for a few years, the ACPC proved unable to deal with the lack of compliance among some members and the growth of supply from non-members, in particular Vietnam, and was disbanded in 2001 at the height of the global coffee crises. Whereas previous, successful attempts at supply management in the twentieth century resulted in a consensus among state officials around the benefits of using coffee statecraft to attain higher coffee prices, the end of the ICA followed by the failure of the ACPC had the opposite effect, turning many of those previously in favor of collective supply management toward accepting the necessity of "free trade."

Mauricio Galindo, current head of operations of the ICO, was directly involved in the ACPC and has experienced declining support for supply management first hand. During an interview in his office at the ICO headquarters in London in March 2013, Galindo highlighted the key dynamics of supply and demand in the global coffee market, before explaining how far the ICO now was from the quota system of the past. Changing geopolitics meant that a quota system was no longer politically feasible and, argued Galindo, markets had ultimately proven to be more economically efficient and transparent than state price regulations. The new role of the ICO would be "to help the market manage supply with a long-term view, with a sustainable view." When I pressed him a bit further about the extent to which governments still have power to manage coffee markets, Galindo replied: "Let's face it, ultimately governments do hold the last card. It's a tough one, it's an extreme one, it's the one you'd never want to flag. But let's face it, in the end, if your government is against you, it doesn't matter who you are . . . they can kick you out."[1]

Galindo's observation reveals that the turn away from state social regulation of coffee markets since the late 1980s is fundamentally not about the decline of state power but about political choices. It reflects the current international consensus among the leaders of the most powerful states around: first, a general refusal to use the tools of statecraft to carry out more radical or robust forms of social regulation, agrarian reform, and human development in the coffee sector; second, a rejection of cooperative behavior among otherwise competing capitalist states on such things as prices, labor rights, and environmental standards in favor of cooperative behavior (through organizations like the WTO) on liberalizing investment and trade and protecting intellectual property; and third, a denial around the continued and pervasive role played by the state in the global coffee market. Competing capitalist states continue to respond to the perceived needs of their territorial and capitalist logics in varied ways, only in a manner that has mostly abandoned the vision of the "developmental state" in favor of a non-developmental state that perpetually denies the ability of the state to act in the broader social interest.

As a result, while a great deal has changed in the coffee industry over the past few decades, many of the overarching patterns described in the previous chapters remain generally the same. Extreme market volatility remains the norm within the industry. Corporate oligarchy remains the dominant feature of the global coffee chain, with Northern-based corporate roasters enhancing their position through ownership of their own trading companies, expanded direct private contracts with giant Southern coffee plantations, and the development of special relationships with major retail supermarket chains. Small coffee farmers and rural workers continue to eke out a precarious existence, highly vulnerable to global market swings and climatic conditions that large-scale coffee plantations and Northern transnational roasters can much more easily adjust to and profit from. The geopolitics of coffee statecraft continues to play a central role in setting the conditions under which these uneven and unjust economic, social, and political outcomes occur and persist, which frequently give rise to diverse forms of social protest, rebellion, and resistance from below. At the same time, long historical patterns within the industry have been impacted by important emerging trends, including the cost-price squeeze, the financialization of coffee, the intensification of environmental crises, and the "rise of the South."

The cost-price squeeze

One important new dynamic in the global coffee industry has been the increasing pervasiveness of the "cost-price squeeze" – when the prices paid for coffee beans are not keeping up with the ever-rising costs associated with growing them. Over the past half-century, TNCs have not only expanded their "downstream" control of processing, retailing, and distributing coffee, but also expanded their "upstream" dominance through the growth of the agro-input industry, which, as Tony Weis has documented, has broken open the traditional "closed-loop agro-ecosystem," where the majority of inputs (such as seed and fertilizer) were attained on-farm or locally. Instead, farming is being transformed into a transnational "through-flow system" where genetically modified seeds and chemical fertilizers, pesticides and insecticides are increasingly purchased from TNCs. Add to this extremely high oil prices, which drive up the costs of transportation and the use of farm machinery, and you have a process that is financially burdensome for medium and small farmers who, once on the corporate treadmill, must continue to purchase costly inputs to remain competitive, leading to growing debt and frequent crises.[2]

The social impact associated with the cost-price squeeze was powerfully demonstrated in February 2013 with the start of a major coffee strike in Colombia, one of the world's top coffee exporters with over 560,000 coffee growers, most of whom are small or medium-sized farmers. Coffee farmers organized a massive strike, blockading major highways and roads, demanding an increase in government subsidies and other supports. Despite relatively high coffee prices at the time, farmers were in crisis due to the rising costs of fertilizer and other imported inputs, made worse by the high currency exchange rate of the Colombian peso. The government initially resisted the farmers' demands, sending out thousands of police to suppress the protestors, tragically killing one and injuring many more. But the strike caused major economic chaos and social instability, threatening both the territorial and capitalist logics of the state. In March, the government gave in, agreeing to $444 million in new spending to boost direct subsidies to farmers, provide new credit, and support the purchase of agro-inputs, along with forgiving all interest and payments associated with 2013 loans to coffee growers from the publicly owned bank, Banco Agrario, which accounts for 90 percent of all national coffee loans. In this instance, coffee's long history of social crisis and social struggle is combined with the new dynamics of an intensified cost-price squeeze.

The financialization of coffee

A second intensifying trend in the global coffee industry in recent years has been the financialization of the global coffee chain, part and parcel of the wider financialization of the entire global food system.[3] The coffee market, due to its volatility and unpredictability, has long been the target of financial speculation. In 1881, the New York Coffee Exchange was incorporated after a disastrous attempt by the three largest US coffee importers, known as the "Trinity," to corner the US market resulted in their bankruptcy, sparking chaos for the entire industry. It was argued that the Exchange would provide risk assurance against price volatility by providing a forum for buyers to contract with a seller to purchase a certain amount of coffee at an agreed time in the future at a guaranteed price; "real" coffee companies would use the contracts as *hedges* against price changes, while speculators would provide *liquidity*. Over time, the Exchange has served merely to escalate price instability as speculators have sought to profit by gambling on or manipulating coffee prices through strategically buying and selling with the immense financial resources at their disposal. From the 1960s to the 1980s, the ICA managed to temper speculative activity in the coffee market by providing more stable prices year after year, which limited opportunities for speculative gains. The ICA's collapse in 1989 ushered in a renewed intensification of speculation and a boom in futures contracts.

More recently, coffee speculation has intensified even further as part of a wider turn toward global commodity speculation. In the wake of the global financial crisis beginning in 2008, after which rich governments bailed out their financial sectors to the tune of trillions of dollars, food commodities in general have become major targets of speculation for the world's largest investment banks: chief among them US-based Goldman Sachs, Morgan Stanley, Citibank, and JP Morgan; British-based Barclays and HSBC; and German-based Deutsche Bank. These activities have driven up prices for most major commodities, increasing corporate profits at the expense of intensifying hunger among the world's poor who cannot afford to meet their basic food needs. According to some estimates, speculative financial investment now controls more than 60 percent of food markets, compared to only 12 percent in 1996.

Food speculation has been allowed to grow, largely unchecked, because it benefits powerful financial companies whose interests are deeply interwoven with the territorial and capitalist logics of the world's richest nations. Peter Gowan, whose notion of "economic statecraft" inspires the use of "coffee statecraft" here, has effectively argued that the nature of the current, highly unpredictable and unstable, global financial architecture emerged historically out of political decisions made by the United States and its allies. Gowan argues that in the 1970s the US government initiated a "Dollar–Wall Street Regime (DWSR)" in response to its declining position in international trade: abandoning dollar–gold convertibility in 1971, allowing it to freely determine the price of the dollar, now entrenched as *the* international currency, in its own interests; and abolishing its capital controls unilaterally in 1974, sparking an inflow of foreign investment into Wall Street and helping to finance the United States' growing trade deficit. Throughout the 1980s and 1990s, states in the North and the South were compelled to remove their own capital controls to remain attractive to investors, initiating a process of "competitive deregulation" and significantly bolstering the power of financial capital and speculators.[4] The recent turn toward intensified coffee speculation is just part of a wider political process based on the further expansion of a chaotic global financial system, one in which powerful, Northern-based investment banks are playing an ever more important role in driving global commodity prices.

The potential pitfalls for the coffee industry of the rapid growth of financial speculation lie in the tendency of financial capital to greatly intensify both price booms and busts when they occur. So far, intensified speculation since 2008 likely helped to further drive up coffee prices until 2011. Prices have since declined and, although markets are extremely difficult to predict, the long history of coffee would suggest that the threat of a bust is ever present. Should this bust occur, the impact of greatly heightened speculation could be disastrous as investors pull out money and further drive *down* prices, having a devastating impact for the most vulnerable in the global coffee market – rural workers and small farmers – who would face layoffs and bankruptcies, while the biggest corporate players – financial banks or coffee companies – would likely survive and possibly even thrive.

Environmental crises

The long history of coffee's integration into the world system and emergence as a major export crop has brought with it numerous environmental impacts. As discussed in previous chapters, in the nineteenth century Brazil, the world's leading coffee grower, pioneered "full-sun," monocrop production, involving removing all forest cover through clear-cutting and then planting coffee trees in rows up and down hills. This method intensifies soil erosion, as well as the depletion of soil nutrients due to the rapid growth of sun-exposed trees, which leads to an escalating reliance on artificial fertilizer.

Monocrop production also has the effect of narrowing down genetic stock due to over-reliance on a limited number of high-producing trees. Today, most of the Arabica beans grown in Latin America derive from only two genetic lines, making Arabicas highly susceptible to a variety of environmental threats, in particular coffee leaf rust and the coffee berry borer worm, both of which continue to have highly destructive impacts throughout the industry.

The introduction of new, "Green Revolution," high-yielding coffee plants and agro-chemicals (pesticides, fertilizers, and insecticides) beginning in the 1970s led to a significant expansion of full-sun, monocrop farming to other regions of the coffee world, and in particular to the most productive and economically efficient coffee countries, beginning with Colombia and Costa Rica and, in more recent years, continuing with other emerging coffee leaders, chief among them Vietnam, which grows nearly all of its coffee full-sun. The turn toward Green Revolution chemicals has not been strictly market-driven and has entailed substantial efforts on the part of state and international agencies, particularly US government agencies and powerful philanthropic organizations such as the Rockefeller Foundation and the Bill & Melinda Gates Foundation. In environmental terms, the result has been an expansion of the destructive patterns associated with full-sun monocropping: increased soil erosion, the destruction of tropical rainforests, and major declines in the biodiversity of trees, birds, animals, and insects. Perhaps most tragically, in many instances vulnerable coffee workers and farmers have been exposed to a variety of chemicals that have been implicated in many illnesses.[5]

While full-sun cultivation has expanded rapidly, however, traditional methods of "shade-grown" coffee have continued to persist on a significant scale. Growing beans under the shade of the existing forest canopy, with the branches trimmed to allow in sunlight, protecting the soil from nutrition loss and erosion, has remained the cultivation method of choice of many small farmers throughout Central America, Ethiopia, Mexico, and other parts of the coffee world who cannot afford the costly, chemical-intensive, full-sun method. The ecological benefits of shade-grown coffee are substantial when compared to full-sun coffee and have been widely documented. The benefits include greater species diversity (one study in Chiapas, Mexico, discovered 180 species of birds in a shade-grown plantation, second only to the diversity of a natural forest); better soil conservation (a study in Venezuela determined that full-sun fields experienced twice the soil erosion of shade-grown); better pest control (one study in Jamaica found a lack of birds to act as natural predators resulted in a 70 percent increase in infestation of coffee berry borer); and less environmental pollution (nitrogen-fixing shade-grown trees reduce the amount of nitrogen released into the atmosphere and water and, by one estimate, potentially reduce the amount of fertilizer a farmer would have to buy by 25–30 percent).[6]

The ecological benefits have often been evoked in defense of the superiority of small versus large-scale coffee farming, when measured in a holistic manner. In fact, even in raw economic terms, despite many myths to the contrary, small farmers tend to be as efficient as large-scale farmers in resource utilization and productivity per unit of area. Beyond this, if we adopt a broader notion of "social efficiency" that takes into consideration environmental and social sustainability, small farmers score even better, generally providing more rural employment than large farmers, relying less on environmentally destructive agro-chemical inputs, and managing the local ecosystem in a more sustainable manner, especially in instances where land is organized communally or cooperatively and local cultural values promoting environmental stewardship remain strong.

Consequently, small-farmer, shade-grown coffee has emerged as a central feature of an array of environmentally sustainable coffee projects – bird friendly, rainforest certified, organic – that have sought to codify, enhance, and promote practices that have long been carried out by coffee farmers. Like fair trade (whose farmers also produce primarily shade-grown beans), these projects have had important impacts and are likely to remain a permanent and growing feature of the global coffee industry. At the same time, the overall reach and breadth have been limited. Organic coffee, for example, has grown rapidly since the early 1980s and continues to do so. And yet, despite this growth, a recent report by the FAO of the UN estimates that certified organic coffee represents only around 2 percent of the EU market and 3 percent of the North American market.[7]

Addressing the wider environmentally destructive impact of the coffee industry on a global scale will ultimately require going beyond where environmental certification alone can take us. Moreover, the greatest environmental threats to the coffee industry come increasingly from forces outside the industry itself: from the impact of climate change caused by the historically unprecedented pace of greenhouse gas emissions leading to global warming. In this context, coffee's impact is significant – in particular, through carbon dioxide (CO_2) emissions associated with the transportation and processing of coffee and agro-chemicals and through the clear-cutting of forests – but just one piece of a much larger problem. The on-the-ground threats posed by climate change to coffee farmers are in many instances similar to those felt by farmers throughout the world; water shortages, for example, caused by increased temperatures and more erratic weather patterns are emerging as major concerns for farmers and development organizations, which seek to promote "adaptation" through new methods of water conservation.

Climate change is also having major impacts specific to the coffee industry. Rising temperatures stress existing coffee trees, making them more vulnerable to pests and disease, and increased average temperatures pose a threat to the genetic diversity of Arabica beans. Recent research conducted by British and Ethiopian researchers has raised the prospect that all of the world's *wild* Arabica beans (as opposed to the standard Arabica beans that are currently grown for consumption) could be wiped out by 2080, due to temperatures rising above the necessary bioclimatic range, destroying a key source of genetic biodiversity that could be beneficial for future industry and non-industry uses. Perhaps most significantly, climate change currently threatens to fundamentally alter the entire growing map for higher-quality Arabica beans, in a manner that is bound to have major social, economic, and political consequences, especially for the poorest and most vulnerable. Estimates suggest that in Central America, a major coffee region known for higher-quality mild Arabica beans coveted by specialty coffee roasters, higher temperatures will push the optimal elevation for coffee higher, ruining the viability of coffee growing in lower regions, while threatening the quality, distinctiveness, and consistency of beans in higher regions. The overall effect will be to significantly shrink the size of specialty coffee growing areas in Central America by 2050, leaving tens of thousands of relatively poor farmers and workers scrambling to find viable new sources of income in lower areas or to respond to the impact of higher temperatures and changes in precipitation patterns in higher ones.[8] Currently, there is little evidence that coffee states and the international community more generally are prepared for or committed to providing the necessary social and economic supports required to significantly aid farmers in these adjustments.

The rise of the South

Perhaps one of the most significant changes in world trade and the global coffee economy in recent years has been the new geopolitics of South–South trade, investment, and cooperation ushered in by what many have termed the "rise of the South." After centuries of uneven development and poorer economic and social advancements than Northern countries, a number of larger Southern countries have experienced decades of substantial economic growth and gradual improvement in most major social indicators. Brazil, Russia, India, and China (which together, combined with the smaller economy of South Africa, form the BRICS group of states) have been at the forefront of new South–South trade relations that are significantly altering world trade patterns. South–South exports now account for nearly half of all Southern exports; South–South foreign direct investment flows grew from $2 billion in 1985 to $60 billion by 2005; and the share of South–South trade of world merchandise trade increased from 8 percent in 1980 to 26 percent by 2011.[9]

These changing economic patterns have been interwoven with new forms of international coordination, regional integration, and socio-political linkages among Southern partners. The results have been new, uneven "hub-and-spoke" relations between Southern countries; the emergence of new instances of Southern competition and cooperation; and the rise of powerful Southern TNCs with global reach and influence – all of which have had highly differentiated repercussions across the South, posing new challenges and opportunities for BRICS countries and smaller Southern states.

In terms of the global coffee economy, the rise of the South has underpinned several of the most significant recent developments. In particular, substantial opportunities for the expansion of coffee exports have emerged out of the growth of major Southern consumer markets, which played significant roles (along with commodity speculation and poor harvests in Colombia) in the 2011 boom in coffee bean prices. According to some estimates, demand for coffee in emerging markets in developing countries could well account for 50 percent of global coffee consumption by 2020. One of these major Southern markets is Brazil, the sixth largest economy in the world. While the country still maintains its title as the world's largest coffee exporter, coffee is much less important to the Brazilian economy than it once was, eclipsed by exports of iron ore, crude oil, soybeans, sugar, and poultry. On the consumption front, domestic coffee consumption in Brazil has been significant since the 1960s, but starting in the mid-1990s it shot up dramatically, especially that of higher-quality Arabica beans. By 2012, Brazil had become the second largest coffee consuming country in the world, after the United States, and could well become first over the next few years given its rate of growth (see figure 6.1). The steady expansion of the Brazilian middle class, leading to millions of consumers with greater disposal income, has underpinned this growth. Their decision to spend their income on coffee, however, as opposed to other products has to an important degree been driven by the strategic decisions of the coffee sector itself. Beginning in the late 1980s, Brazil's private coffee roasters, organized into the Brazilian Coffee Roasters' Association (Associação Brasileira da Indústria de Café, or ABIC), launched a multimillion-dollar marketing campaign to promote domestic coffee consumption through television advertisements, celebrity endorsements, and efforts to tighten quality standards for Brazilian beans. From 1989 to 1998, ABIC spent $26.7 million dollars on this campaign, during which time total coffee consumption in Brazil increased by nearly 65 percent.[10]

The most novel coffee markets, however, have emerged in Asia, where countries generally have not had significant

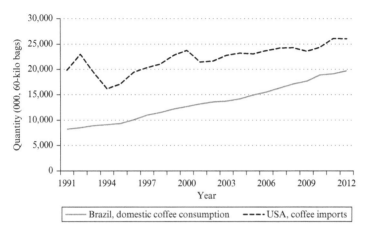

Source: ICO statistical database (http://www.ico.org), accessed July 30, 2013.

Figure 6.1   Brazilian coffee consumption vs. US coffee imports, 1991–2012.

coffee consuming traditions. Several Asian economies have been particularly central to the rise of the South, experiencing rapid and sustained economic growth and accounting for the vast majority of global poverty reduction since the 1980s, measured by the UN and other official institutions. Chief among them has been China, whose economic success has involved a model far removed from free trade fantasies, entailing massive state involvement in the economy, regulations on foreign investment and capital, and gradual trade liberalization. From 1981 to 2005, the poverty rate in China declined from 85 percent to 15 percent, fuelling the fastest growing consumer market in the world, with China becoming the world's largest purchaser of everything from cars to timber, gold to pork, and possibly soon coffee. While there are other Asian economies that have larger coffee markets, in particular South Korea, coffee sales in China have soared, increasing by around 15 to 20 percent annually from 2006 to 2012 (compared to the world average of 2 percent), and China is set to become one of the largest coffee markets in the world by 2020.[11]

New opportunities for market growth, however, have also come with new and intensified competition among coffee producers and exporters. The "rise of the South" is highly uneven, with some countries and regions doing much better than others. Not all Southern countries have been magnets for trade and investment, or experienced rapid economic growth and substantial increases in their political weight. Nor have all people within the South been benefiting from increased trade and investment, due to major inequalities within and between Southern nations.

In the coffee industry, unevenness between Southern countries is reflected in the rise of several Asian coffee exporters, matched by the decline in the African coffee exporters that had dominated Robusta coffee markets from the 1950s to the 1980s. Whereas most African coffee countries in the 1980s and 1990s, under intense international and domestic pressure, pursued neoliberal statecraft devoted to trade liberalization and privatization, Asian coffee exporters, to varying degrees and in different ways, pursued more gradual privatization along with more robust dirigiste roles for the state in direct investment in coffee through SOEs, publicly funding research, and the provision of various agriculture extension services and infrastructure. Today, three of the world's top six coffee exporters are Asian countries – Vietnam, Indonesia, and India – and the region exports over three and a half times more coffee than Africa (see figure 6.2). Politicians and policy makers in Africa are not unaware of this displacement and its negative impacts on the coffee sector and African economies in general. In a statement read on behalf of the Inter-African Coffee Organization (IACO) at the ICO meetings in March 2013, Aly Touré, representative of Côte d'Ivoire, declared:

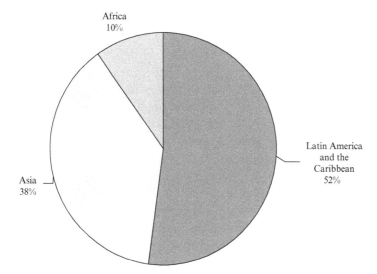

Source: ICO statistical database (http://www.ico.org), accessed November 14, 2013.

Figure 6.2 Coffee exports by region, 2012.

Ladies and Gentlemen, the story of African coffee is well known to all of us. Africa had generally declined, in its influence in the global coffee dynamics, following the liberalization of the sector in our respective countries, which came on the heels of the collapse of the ICO quota system. In many countries, owing to the pursuit of policies of structural adjustment, and the declining role of governments in the support to the coffee sector, farmers' earnings plummeted as a result of poor prices, resulting in the abandonment of coffee farms. Subsequently, production declined and quality worsened.

Thus, while in 1970 Africa contributed about 32% of global coffee production, today the continent only produces about 12%. The ramifications of these developments to us as producers, and as sovereign states, have been devastating, especially for countries that derived significant foreign earnings from coffee exports. In many instances, reduced foreign exchange negatively affected national development programmes.[12]

To overcome "Africa's bleak coffee history," the statement declared the commitment of African states to work collectively toward a new "renaissance" in African coffee, with a renewed emphasis on promoting productivity, quality, and sustainability. The ways and means of achieving these goals were not directly addressed, although the representative of Tanzania proposed that a new "logic of intervention" would be required to do so.[13] For those poorer and more vulnerable states on the margins of the rise of the South, while they are not attaining the lion's share of the benefits, it seems that new policy space, ideological frames, and political confidence are emerging, allowing for a rethinking of coffee statecraft and a push toward new, alternative forms of coffee craft that are often subtle, but still significant.

On the margins of the "rise of the South"
Perhaps one of the most outstanding examples of coffee statecraft at the margins in recent years has been in Ethiopia, home to some of the world's top-quality Arabica beans, as well as one of the poorest countries in the world. While Ethiopian coffee exports represent less than 3 percent of the world's total, coffee accounts for over 31 percent of the value of all Ethiopian exports – making Ethiopia highly dependent on a global industry in which it has limited economic or political weight.

Beginning in 2005, the Ethiopian government has sought to address its tenuous position in the global coffee economy through a new approach to managing its coffee statecraft. The first step involved a campaign to trademark its most renowned coffee beans, Sidamo, Harar, and Yirgacheffe. As Ethiopia was one of the birthplaces of coffee, the government claimed that the country and its farmers should have intellectual property rights over the use of the specialty coffee brands and applied for the trademark registrations in major coffee consuming nations. As mentioned in chapter 5, the trademarking campaign, with the support of Oxfam International, initially had to overcome resistance from the NCA and major coffee companies before successfully registering its trademarks in the US in 2007, followed by around 30 countries in the years that followed.

As a result of the trademarking initiative, the Ethiopian Intellectual Property Office (EIPO) of the Ethiopian government has become the legal owner of Sidamo, Harar, and Yirgacheffe trademarks. In consultation with the Ethiopian Fine Coffee Stakeholder Committee – composed of Ethiopian exporters, cooperatives, and government agencies – EIPO manages the trademarks and the Ethiopian "brand," which only licensed companies can use. In sharp contrast to most private trademarking and branding initiatives, however, being licensed to use Ethiopian coffee does *not* require a royalty fee – something that Ethiopian officials no doubt feared would chase companies away from buying Ethiopian beans. Instead, licensed companies agree to explicitly market and promote Ethiopian brands in various ways. The trademarking initiative was followed in 2008 by the creation of the Ethiopian Commodity Exchange (ECX), Africa's first commodity exchange, as a jointly owned, public–private company, through which most of the country's coffee, and many other commodities, must now be traded. The ECX has been designed to make exchanges between buyers and sellers more reliable and transparent by disseminating market information; coordinating delivery, payment, and various forms of "risk management" (such as future contracts); and overseeing and guaranteeing product quality and quantity.

Taken together, the trademarking initiative and the creation of the ECX represent an attempt by the Ethiopian government to use market-friendly coffee statecraft to enhance the symbolic and quality value of Ethiopian coffee and gain a greater share of that value for Ethiopian farmers. Daviron and Ponte have argued that market power in the coffee industry comes not just from capturing the material value of the coffee beans, but also from the ability to define the coffee identity, norms, quality standards, and symbolic attributes. The trademarking initiative and the creation of the ECX represent a struggle to gain greater control over the non-material, subjective elements embedded in Ethiopian coffee beans – its quality, geographic, and symbolic attributes – and use that control to direct the extra value that these attributes can garner on Northern markets into the hands of Ethiopians.[14]

It is too early to gauge the overall success of the trademarking and ECX initiatives, which are devoted to longer-term goals that take time to play out. One weakness that is already apparent, however, is that it is not clear how much extra value they are actually creating. Light Years IP, a non-profit organization that assisted Ethiopia with its trademarking efforts, claims that Ethiopian coffee farmers' income doubled, with an additional $100 million in 2007–8, but it is not clear how much of this was driven by a jump in international coffee bean prices that year.[15] The ECX provides greater transparency, reliability, and quality control for Ethiopian beans, but it is also not clear how much extra value is actually being created and how much is finding its way to poor coffee farmers and workers. Ethiopian coffee prices have continued to follow international prices in much the same way as beans from other exporters, and small coffee farmers and workers have continued to live off very low incomes. At the same time, whereas most African countries have experienced major declines in coffee production and export since the start of the millennium, Ethiopia has bucked this trend; its share of global coffee exports has modestly increased and Ethiopia is now Africa's largest and the world's tenth largest coffee exporter overall.

In the end, Ethiopia's particular form of coffee statecraft would appear to be having a moderate impact on the global distribution of value along the coffee chain. And yet, the limitations of Ethiopian statecraft must be taken into consideration, with its relatively vulnerable position in the coffee industry, the global economy, and the geopolitical map ruled by more powerful states. The trademarking and ECX initiative both reflect growing recognition in Ethiopia of the benefits of more consciously and directly applying coffee statecraft toward developmental ends, and the Ethiopian state had to wage fierce battles to win the right to carry them out. It is part of a subtle, but still significant, turn away from the dictates of the free trade package and toward a new "logic of intervention" among African coffee countries, one based on an emerging Southern consciousness and political confidence and inspired by the complex and contradictory success stories of new Asian coffee leaders.

The rise of fair trade South

An emerging Southern consciousness and political confidence are apparent not only in Southern statecraft, but also among social movements within the coffee sector organized by poor and marginalized farmers, rural workers, and rural landless. As discussed in chapter 2, states do not simply pursue their territorial and capital logics in a political and social vacuum, but rather are driven and limited by social pressures stemming from a perpetual struggle between dominant and subordinate groups. The coffee fields have always been fertile grounds for revolution, rebellion, and revolt against powerful coffee elites and Northern imperialist states. In recent years, more subtle forms of Northern dominance have also become targets of resistance "from below" – including progressive-minded movements like fair trade coffee.

Despite genuine concern for the needs of small farmers and

rural workers, fair trade certification has long been critiqued for allowing Southern participants only a modest or even token role in developing, managing, and overseeing fair trade rules and regulations. FLO, the world's top fair trade umbrella organization (discussed in chapter 5), was originally founded in 1997 composed of 17 Northern national fair trade initiatives (19 today), which dominated the General Assembly and Board of Directors, with only a single, non-voting producer representative on the Board. Over the years, under intense pressure from Southern producers, FLO has reformed its structure, allowing a more even distribution between producer and consumer representatives on the Board in 2005, and then reformulating the General Assembly in 2011 to constitute 50 percent producer representation, as opposed to the previous 14 percent.

A key organization in pressing for these changes has been CLAC, founded in 2005 by fair trade farmers seeking a greater voice in the fair trade network.[16] Headquartered in El Salvador, CLAC currently represents around 300,000 small producers from organizations in 21 countries. Its members grow a variety of agricultural products, of which coffee is the most significant. CLAC is an official representative of Latin American and Caribbean fair trade producers at the FLO General Assembly and since its founding has engaged in numerous activities within and outside the FLO system, including launching the Foundation of Organized Small Producers (Fundación de Pequeños Productores Organizados, or Fundeppo) in 2010 to manage CLAC's own certification system based on the Small Producer's Symbol (SPS).

While compatible with the overall model of product labeling and ethical certification, CLAC's vision is different and broader than FLO's in several ways, three of which are perhaps the most significant. First, CLAC is driven by and explicitly devoted to small farmers and is firmly opposed to the gradual opening of fair trade to giant coffee plantations. CLAC argues that small farmers, organized into cooperatives, represent an alternative and more socially and ecologically just model to corporate coffee plantations that already have significant economic and political advantages over small farmers in the global market.

Second, whereas the coffee value chain in fair trade flows South-to-North along the same lines as conventional coffee chains, CLAC is explicitly devoted to fostering domestic markets and South–South fair trade to diversify away from over-reliance on Northern markets. Thus far, market realities have compelled CLAC members to remain dependent on the North for the majority of their income, but several Southern fair trade bodies have emerged in recent years and the movement is growing. CLAC's own SPS initiative seeks to address both of these concerns, certifying only small farmers and targeting Southern markets as much as possible.

Third, whereas FLO limits its advocacy efforts to issues directly related to fair trade labeling, CLAC has much more ambitious political and advocacy agendas that go beyond ethical consumerism. These entail support for global movements promoting cooperatives and "food sovereignty." Whereas the dominant notion of "food security" is premised on global market integration into a food system driven by TNCs, the growing "food sovereignty" movement promotes national self-sufficiency based on local farm production; an emphasis on healthy, good-quality, "culturally appropriate" foods; and a political agenda dedicated to a more just balance between the agricultural and industrial sectors, land reform and land redistribution, discouraging chemical-intensive methods and biotechnology, valorizing peasant and indigenous agricultural wisdom, and promoting small farmer, cooperative, or state enterprise agricultural production.

CLAC's political agenda is also more directly concerned than fair trade North with the state and the ways that statecraft can inhibit or support the livelihoods of small farmers. CLAC is particularly opposed to the general decline of state support for small farmers in the neoliberal era, even while states continue to significantly favor plantation owners' access to state-provided credit, subsidies, and infrastructure. Consequently, CLAC has shown cautious optimism for the rise of numerous and diverse social democratic governments in Latin America and the Caribbean since the early 2000s, as well as an interest in collaborative efforts between state institutions and social movements that have resulted in government legislation explicitly promising financial and planning support to develop the national fair trade market, alongside "solidarity economies" and organic markets, in Brazil (2010), Bolivia (2011), and Ecuador (2008 and 2011).

These state initiatives, like CLAC's own independent efforts to promote its SPS certification for small farmers, are incipient and have not yet sparked a significant surge in Southern fair trade markets. They do, however, reflect important changes that are underway related to coffee and other global commodities, rooted in the growing economic and political weight of Southern economies and new Southern-focused and South–South efforts between grassroots movements, policy makers, and political and economic elites. The full impact of these changes on the global coffee economy in 10 or 20 years remains to be seen, but they are likely to be significant. Perhaps above all, they suggest that in the future the influence of Southern countries on the consumer side of the global coffee chain – from buying to branding to governing conventional, fair trade, organic, or other ethical markets – is bound to be much greater than has historically been the case.

Coffee crises . . . again

Despite the positive emergence of an organization such as CLAC, despite signs that the dominant agenda of the free trade package is increasingly being challenged and questioned internationally, and despite the appearance of new Southern trade dynamics, many of the dominant structures of the global coffee industry remain firmly entrenched; the coffee economy continues to be highly volatile, grossly unequal, ecologically destructive, and, of course, immensely profitable for a lucky minority of wealthy investors, plantation owners, and state elites. Far removed from the luxurious board rooms of Kraft Foods Group, Nestlé, and Starbucks, the coffee industry today exists in a state of daily crisis for the poorest workers and farmers, who often cannot afford access to basic food, proper housing, essential health care, and schooling for their children. These daily crises periodically erupt into major disasters.

One such disaster emerged as I was writing this book. Beginning in 2012, a coffee leaf rust outbreak hit Latin America, having the worst effect on the main coffee exporting countries of Central America (Costa Rica, El Salvador, Guatemala, Honduras, and Nicaragua). Central American coffee countries account for around 12 percent of world production, and are particularly important in providing high-quality Arabica beans. The rust outbreak is the worst that the region has seen since 1976, when rust first appeared in Central America. According to the ICO, for the 2012/13 crop year, over 50 percent of the total coffee growing area in the region was affected, causing a drop of nearly 20 percent in coffee production and a loss of approximately $500 million. Millions of coffee farmers, the majority of whom are small-scale in Central America, experienced significant drops in income, while vulnerable rural workers were faced with mass layoffs: approximately 374,000 coffee jobs in the region were lost, representing over 17 percent of the entire coffee growing workforce. Some estimate that regional production could drop even further by the end of the crop year in 2014, by as much as 50 percent, which could have devastating social, economic, and political impacts throughout the region.

The Central American coffee crisis and its roots are inseparable from the historical and global dynamics of the coffee industry discussed in this book. A major factor in the rapid and extensive outbreak of coffee leaf rust has been the spread since the 1970s of full-sun, monocrop cultivation, which now covers approximately 40 percent of the coffee areas of Central America. This form of cultivation has substantially reduced the genetic stock of Arabica trees, as well as the biodiversity of natural predators of coffee pests, making Central American trees exceedingly vulnerable to the spread of disease. As is always the case in the coffee world, the negative impacts of the crisis will be divided highly unequally along class lines. Even while most full-sun cultivation has been carried out by the region's wealthiest farmers, rural workers and small farmers (who grow the vast majority of their coffee under far more environmentally sustainable shade-grown conditions) will pay the greatest price, as they lack the wealth and resources to buffer themselves from the worst effects of the crisis.[17]

The impact of the coffee crisis will also be uneven internationally. Over the longer term, the crisis threatens to hasten the ecological impacts of climate change on the prime growing regions in Central America. Consequently, leading representatives of Central American nations and specialty coffee companies at the 2013 ICO meeting voiced deep concerns that the crisis might last several years and bring about the end of the region's comparative advantage based on higher-quality, and higher-priced, Arabica beans. In place of Central American Arabicas, it is likely that coffee roasters will turn increasingly to Robusta beans with new processing technologies applied that soften Robustas' otherwise harsh taste. Thus, as Central American coffee farmers confront the possibility of catastrophe, Robusta coffee leaders, like Vietnam and other Asian and African exporters, could well be facing continued expansion.

These outcomes do not stem strictly from a combination of nature and markets – supply and demand dovetailing with ecological crises – but are deeply interwoven with coffee state-craft and the role that states play in managing the conditions under which coffee is grown, traded, sold, and consumed. As is so often the case in these neoliberal times, coffee states have stepped forward to oversee a desperate bailout after the crisis has already begun: governments in Costa Rica, Guatemala, and Honduras have declared phyto-sanitary emergencies; all Central American countries have pledged tens of millions of dollars to assist their domestic growers; and the ICO has coordinated an international campaign seeking hundreds of millions of dollars in loans and grants from multilateral financial institutions, international donors, the private sector, and government development agencies to support special pesticide applications targeting rust; fund tree renovation programs to replace old coffee trees with newer, more rust-resistant ones; and provide farmers with soft loans and various food security and crop diversification programs to partially offset declining incomes. It remains to be seen how much support will ultimately materialize and how effective it will be. What is certain is that hundreds of thousands of rural workers have already lost their jobs and millions of small farmers are already confronting crises, while after-the-fact support such as this disproportionately helps those larger and wealthier farmers who are already able to withstand disasters and can remain standing long enough to see aid actually emerge.

Coffee statecraft's role in this crisis far precedes the bail-out mode of current governments. The uneven impacts and outcomes of the crisis stem from an unequal and unjust global coffee economy forged through the actions of imperialist and autocratic states over hundreds of years of colonialism, slavery, and the expansion of a capitalist world system. The extreme volatility and unpredictability of the global coffee market stem from political and economic elites choosing to meet the territorial and capitalist logics of their states through intensified competition rather than through enhanced collective action of the sort pioneered by the ICA. This decline in collective action has been paralleled by declining commitments by states toward legally mandated, state-managed forms of social regulation on the domestic scene. Instead, they have ceded this terrain to private, market-driven, voluntary initiatives like fair trade and CSR, which can never match the state's reach and impact, and cannot address the root causes of the coffee crisis. Having denied that the state has a robust role to play in managing the coffee economy, states will now be dragged further into managing the crisis to protect their own stability in the wake of mounting social, economic, and political tensions. We can only hope this role will be to provide further social protection and provision for the poor, but the possibility is always there for states to become more militarized and brutal in clamping down on social protest and dissent – something sadly familiar to the past and present of much of Central America.

In seeking to address coffee crises – whether the "big" one in Central America today or the everyday crises of coffee faced by the poorest workers and farmers throughout the globe – the main issue is not whether or not the state should intervene in the coffee market. Instead, we must break free from the intervention/non-intervention trap, challenge and subvert the dominant understandings of the free trade package, and place far greater emphasis on recognizing the political roots of what are so often seen as economic forces at work. What is required is a greater push for *better* coffee statecraft, guided by the history of gains and losses in the highly imperfect global coffee market. This better coffee statecraft should include, first, a much greater emphasis on states directly supporting smaller and medium-sized coffee farms, and a more equitable distribution of land, infrastructure, and resources – something that the majority of the world's most successful coffee industries have done in highly varied ways, from Costa Rica to Colombia to Vietnam. Second, better coffee statecraft requires a recognition of the benefits of collective action among all coffee states to manage markets for greater stability and higher incomes, building on the example of the ICA. This is a lesson that, unfortunately, must perpetually be relearned by states: highly competitive Costa Rica, for example, abandoned the ICA in 1989, only to then join the failed ACPC cartel four years later when global coffee prices began their downward spiral. Finally, better coffee statecraft must involve strong state support for more environmentally sustainable methods of coffee growing, such as that pioneered by small Arabica growers, that can both limit the coffee industry's contribution to climate change and help mitigate its negative effects. Thus far, there is limited evidence of coffee states adopting such a role in any substantive manner, but organizations like CLAC and other social movements are tirelessly engaged in the struggle to change this.

Conclusion

The goal of better coffee statecraft might seem like an elusive one, especially as so many coffee states seem top-down, hierarchical, unresponsive, and in some instances autocratic. Moreover, state leaders and policy makers are often the first to disavow any ability to control or manage markets effectively. They must, however, not be taken at their word. The long history of coffee is replete with examples of farmers and workers demanding and getting more out of their states than they would otherwise offer. Recall that in Colombia in February 2013, when hundreds of thousands of coffee farmers went on strike, the government fervently rejected farmers' demands and brutally sought to suppress the protests; less than two weeks later, however, the government abruptly changed its tune, agreeing to hundreds of millions of dollars of new subsidies and supports. Following this, in August 2013, Brazil, fearing the potential political and social impacts of declining coffee prices, announced plans to purchase as many as three million bags of coffee to prop up prices.[18]

Coffee states have played and continue to play a central role in protecting – and in some cases opposing – the unequal distribution of coffee land and resources, in ensuring a specific set of social relations around commodity production for export, and in the separation of the economic and political spheres, a central characteristic of a capitalist economy. Consequently, the economic and political inequality in the coffee industry cannot be resolved through market adjustments, but rather requires attention to the ways in which coffee statecraft has forged these deep structural roots. Long-term, substantive improvements in the coffee world will require engaging in political struggles to change the direction of coffee statecraft, resisting the allure of quick fixes, and, as Ilan Kapoor has observed, embracing "long-term democratic struggles, with all the risks and setbacks that they entail (i.e. their success is never guaranteed)."[19] If, in the end, the state "holds the cards," then the majority of the world's coffee producers need those cards played much better. Otherwise, the daily crises of coffee will remain for now and the foreseeable future.

Notes

1 The ICO quote comes from ICO, *Monthly Coffee Market Report: May 2013* (London: ICO, 2013). The fear is repeated in future reports, including for October 2013. The reporters' quote comes from Alexandra Wexler, Jeffrey T. Lewis, and Leslie Josephs, "Brazil Drought Jolts Commodities' Prices," *Wall Street Journal* (March 4, 2014), http://online.wsj.com/news/articles/SB1000 14240527023045850045794194736544460330, accessed March 5, 2014. Unless otherwise indicated, all prices in this book are in US dollars. The composite indicator prices used throughout are based on the "1976 version" drawn from the United Nations Conference on Trade and Development (UNCTAD) statistical database (http://unctadstat.unctad.org), which allows for charting prices back to 1960, essential for making historical comparisons. The ICO website (http://www.ico.org) is an excellent source on coffee data, including daily price updates. The calculations regarding real value prices during the crisis come from Oxfam International, *Mugged: Poverty in Your Coffee Cup* (Boston and Washington, DC: Oxfam International, 2002).

2 ICO, *Burundi: Country Datasheet* (London: ICO, 2011). See the online statistical databases of the World Bank (http://data.worldbank.org) and the United Nations Development Program (UNDP), Human Development Reports (http://hdr.undp.org/en/statistics).

3 Oxfam International, *Mugged*, pp. 22–5.

4 For global activists, see Oxfam International, *Mugged*, and the websites of Global Exchange (www.globalexchange.org) and the Fairtrade Foundation (www.fairtrade.org.uk). For more on the historical and current orientation of trade justice activists,

see Gavin Fridell, *Fair Trade Coffee: The Prospects and Pitfalls of Market-Driven Social Justice* (Toronto: University of Toronto Press, 2007); Daniel Jaffee, *Brewing Justice: Fair Trade Coffee, Sustainability, and Survival* (Berkeley: University of California Press, 2007); Ian Hudson, Mark Hudson, and Mara Fridell, *Fair Trade, Sustainability and Social Change* (New York: Palgrave Macmillan, 2013).

5 These quotes from fair trade advocates come, in the order in which they are presented, from: Fair Trade USA, "Vision Statement," http://www.fairtradeusa.org/about-fair-trade-usa/mission, accessed November 4, 2013; Fairtrade International, *Unlocking the Power: Annual Report 2012–13* (Bonn: Fairtrade International, 2013), p. 5; Sean McHugh, executive director of the Canadian Fair Trade Network, "The Importance of Trade," *Fair Trade: Canada's Voice for Social Sustainability* 2 (Summer/Fall 2013), p. 29.

6 See the Coffee Reporter, "NCA Joins Industry Leaders Addressing World Crisis Solutions at Sentercafe," *Coffee Reporter (National Coffee Association of USA)* 8: 12 (2003); Brink Lindsey, *"Fair Trade" and the Coffee Crisis* (London: Adam Smith Institute, 2004).

7 For example, see the World Bank's Sustainable Coffee Landscape Project in Burundi: World Bank, *Project Information Document (Appraisal Stage) – Sustainable Coffee Landscape Project – P127258 (English)* (February 8, 2013), p. 3, http://data.worldbank.org, accessed November 3, 2013. The project has been heavily critiqued by the UN special rapporteur on the right to food, Olivier De Schutter, for pressuring for full privatization despite negative and volatile impacts on small growers. See the press release by De Schutter, "World Bank-Led Privatization of Burundian Coffee Industry Must Not Repeat Errors of the Past" (April 18, 2013), http://www.srfood.org/en/news, accessed November 3, 2013.

8 See My Virtual Coffeehouse, "For the Good Earth," http://myvirtualcoffeehouse.com/for-the-good-earth, accessed November 3, 2013. My Virtual Coffeehouse is a website run by the National Coffee Association (NCA) in the United States.

9 For works that deal with the historical evolution of coffee see Anthony Winson, *Coffee and Democracy in Modern Costa Rica* (Toronto: Between the Lines, 1989); David McCreery, *Rural*

Guatemala, 1760-1940 (Stanford, CA: Stanford University Press,
1994); Robert G. Williams, *States and Social Evolution: Coffee
and the Rise of National Governments in Central America* (Chapel
Hill: University of North Carolina Press, 1994); Jeffery M. Paige,
*Coffee and Power: Revolution and the Rise of Democracy in Central
America* (Cambridge, MA: Harvard University Press, 1997);
William Gervase Clarence-Smith and Steven Topik (eds.), *The
Global Coffee Economy in Africa, Asia, and Latin America, 1500–
1989* (New York: Cambridge University Press, 2003). For works
devoted to power, politics, and class, see John M. Talbot, *Grounds
for Agreement: The Political Economy of the Coffee Commodity
Chain* (Oxford: Rowman & Littlefield, 2004); Christopher Bacon,
"Confronting the Coffee Crisis: Can Fair Trade, Organic, and
Specialty Coffees Reduce Small-Scale Farmer Vulnerability in
Northern Nicaragua?" *World Development* 33: 3 (2005); Benoit
Daviron and Stefano Ponte, *The Coffee Paradox: Global Markets,
Commodity Trade and the Elusive Promise of Development* (London:
Zed Books, 2005); Roldan Muradian and Wim Pelupessy,
"Governing the Coffee Chain: The Role of Voluntary Regulatory
Systems," *World Development* 33: 12 (2005); Jytte Agergaard
Larsen, Niels Fold, and Katherine Gough, "Global–Local
Interactions: Socioeconomic and Spatial Dynamics in Vietnam's
Coffee Frontier," *Geographical Journal* 175: 2 (2009); Jeffrey
Neilson and Bill Pritchard, *Value Chain Struggles: Institutions
and Governance in the Plantation Districts of South India* (Oxford:
Wiley-Blackwell, 2009). Insightful journalistic works include
Anthony Wild, *Coffee: A Dark History* (New York: W. W. Norton,
2005); Gregory Dicum and Nina Luttinger, *The Coffee Book:
Anatomy of an Industry from Crop to the Last Drop*, rev. edn. (New
York: New Press, 2006); Mark Pendergrast, *Uncommon Grounds:
The History of Coffee and How It Transformed Our World*, 2nd edn.
(New York: Basic Books, 2010). Most of these works have little to
say about the geopolitics of coffee statecraft or tend to replicate
the distinction between state and market being critiqued here.

10 The quote is from Robert H. Bates, "Institutions and
Development: Talk for the First International Coffee
Organization World Coffee Conference" (London: ICO, 2001),
p. 13. See Bates, *Open-Economy Politics: The Political Economy of
the World Coffee Trade* (Princeton, NJ: Princeton University Press,
1997). Bates is a lead institutionalist thinker. For a critique of this

approach, see Vivek Chibber, "Building a Developmental State: The Korean Case Reconsidered," *Politics and Society* 27: 3 (1999).

11 Talbot, *Grounds for Agreement*. For more on the global value chain approach, see Daviron and Ponte, *Coffee Paradox*; Niels Fold and Bill Pritchard, *Cross-Continental Food Chains* (London: Routledge, 2005); Henry Bernstein and Liam Campling, "Commodity Studies and Commodity Fetishism I: *Trading Down*," *Journal of Agrarian Change* 6: 2 (2006); "Commodity Studies and Commodity Fetishism II: 'Profits with Principles'?" *Journal of Agrarian Change* 6: 3 (2006); Jennifer Bair (ed.), *Frontiers in Commodity Chain Research* (Stanford, CA: Stanford University Press, 2009); John M. Talbot, "The Comparative Advantages of Tropical Commodity Chain Analysis," in Bair, *Frontiers of Commodity Chain Research*.

12 See Bernstein and Campling, "Commodity Studies and Commodity Fetishism I," and "Commodity Studies and Commodity Fetishism II"; Talbot, "Comparative Advantages." In addition, as noted by Talbot, "Comparative Advantages," the focus of global value chain literature on developing specific typologies for different chains has suffered from the weakness that it is difficult, if not impossible, to define an entire chain on the basis of a single typology: different firms may dominate along different nodes of a chain, and this can change significantly over time. Concepts such as "lead firms" and "drivenness" are, thus, highly instructive, although in this work I employ them in a manner looser than is typically the case in much global value chain work.

13 See David Harvey, *The New Imperialism* (Oxford: Oxford University Press, 2003); Peter Gowan, *The Global Gamble: Washington's Faustian Bid for World Dominance* (London: Verso, 1999); Ellen Meiksins Wood, *Empire of Capital* (London: Verso, 2005); Giovanni Arrighi, *Adam Smith in Beijing: Lineages of the Twenty-First Century* (London: Verso, 2007). These approaches are devoted to understanding imperial powers, but can be instructively applied to all capitalist states. Harvey's work on the two logics of capitalism is based on a reformulation of Arrighi's earlier conception. Peter Gowan's unique use of Marxist political economy in dialogue with neo-realist international relations theory is central to the notion of "economic statecraft," as is Ellen Meiksins Wood's insight into the manner in which capitalist

states work to preserve the artificial separation between the economic and political realms. The historical work of Steven Topik, which emphasized Southern agency in the coffee world, has also indirectly influenced the notion of coffee statecraft. See Clarence-Smith and Topik, *Global Coffee Economy in Africa, Asia, and Latin America*.

14 Wood, *Empire of Capital*.

15 David McNally, *Against the Market: Political Economy, Market Socialism and the Marxist Critique* (London: Verso, 1993), pp. 162–7; Peter Gibbon and Stefano Ponte, *Trading Down: Africa, Value Chains, and the Global Economy* (Philadelphia: Temple University Press, 2005); Duncan Green, "Conspiracy of Silence: Old and New Directions on Commodities," paper presented at the Strategic Dialogue on Commodities, Trade, Poverty and Sustainable Development Conference, Faculty of Law, Barcelona (June 13–15, 2005).

16 Talbot, *Grounds for Agreement*.

17 Pendergrast, *Uncommon Grounds*.

18 Statistics on Nestlé are taken from *Fortune Magazine*, "Global 500" for 2012, at *CNN Money* online, http://money.cnn.com/magazines/fortune/global500/2012/snapshots/6126.html, accessed November 26, 2012. Statistics on Vietnam are based on 2011 figures, taken from the World Bank data online, http://data.worldbank.org/country/vietnam, accessed November 26, 2012.

19 For more on specialty coffee see Pendergrast, *Uncommon Grounds*. For corporate advertising see Michael Dawson, *The Consumer Trap: Big Business Marketing in American Life* (Urbana: University of Illinois Press, 2003). Starbucks' total net revenues for fiscal year 2013 were $14.9 billion, of which 74 percent ($11 billion) came from stores in the Americas. See Starbucks, *Fiscal 2013 Annual Report* (Starbucks, 2013). In 2012, the retail value of the entire US market was estimated at around $30–2 billion, with the specialty coffee industry accounting for a 37 percent volume and a 50 percent value share. See Specialty Coffee Association of America, "Specialty Coffee Facts & Figures" (March 2012), http://www.scaa.org/PDF/resources/facts-and-figures.pdf, accessed August 29, 2012.

20 Dimitris Milonakis and Ben Fine, *From Political Economy to Economics: Method, the Social and the Historical in the Evolution of Economic Theory* (London: Routledge, 2009).

21 Jeffrey D. Sachs, *The End of Poverty: Economic Possibilities of Our Time* (New York: Penguin, 2005); Joseph Stiglitz and Andrew Charlton, *Fair Trade for All: How Trade Can Promote Development* (New York: Oxford University Press, 2006); Jagdish Bhagwati, *Termites in the Trading System: How Preferential Agreements Undermine Free Trade* (Oxford: Oxford University Press, 2008); Paul R. Krugman, *The Conscience of a Liberal* (New York: W. W. Norton, 2009).

22 The notion of free trade as a political, economic, and ideologically charged "package" is derived from Janice Peck, *The Age of Oprah: Cultural Icon for the Neoliberal Era* (Boulder: Paradigm, 2008), p. 8.

23 Jodi Dean, *Democracy and Other Neoliberal Fantasies: Communicative Capitalism and Left Politics* (Durham, NC: Duke University Press, 2009), pp. 50, 55–8.

24 Louis Lefeber and Thomas Vietorisz, "The Meaning of Social Efficiency," *Review of Political Economy* 19: 2 (2007).

25 See Starbucks, *2012 Global Responsibility Report: Year in Review* (Starbucks, 2012), p. 4.

CHAPTER 2

1 Ellen Meiksins Wood, *The Origin of Capitalism* (New York: Monthly Review Press, 1999). See also Karl Polanyi, *The Great Transformation: The Political and Economic Origins of Our Time* (Boston: Beacon Press, 1944); Robert Brenner, "The Social Basis of Economic Development," in J. Roemer (ed.), *Analytical Marxism* (Cambridge: Cambridge University Press, 1985); David McNally, *Another World Is Possible: Globalization and Anti-Capitalism* (Winnipeg: Arbeiter Ring, 2006).

2 Sachs, *End of Poverty*, pp. 31, 50.

3 For more on coffee's early expansion, see Steven C. Topik, "Coffee," in Steven C. Topik and Allen Wells (eds.), *The Second Conquest of Latin America: Coffee, Henequen, and Oil During the Export Boom* (Austin: University of Texas Press, 1998); Steven Topik, "The Integration of the World Coffee Market," in William Gervase Clarence-Smith and Steven Topik (eds.), *The Global Coffee Economy in Africa, Asia, and Latin America, 1500–1989* (Cambridge: Cambridge University Press, 2003).

4 The history of the world system in this chapter relies on Polanyi, *Great Transformation*; Brenner, "Social Basis of Economic Development"; Eric R. Wolf, *Europe and the People without History* (Berkeley: University of California Press, 1997); Wood, *Origin of Capitalism* and *Empire of Capital*. Much of the general historical narrative of global coffee, as well as statistics on prices, consumption, labor movement, and production during the colonial period, is drawn from the rich empirical work of Pendergrast, *Uncommon Grounds*. An earlier chapter on the history of coffee appeared in Fridell, *Fair Trade Coffee*, pp. 101–34.

5 Sidney W. Mintz, *Sweetness and Power: The Place of Sugar in Modern History* (New York: Penguin, 1985), pp. 19–74; Wolf, *Europe and the People without History*, pp. 195–201.

6 Brenner, "Social Basis of Economic Development"; Wood, *Origin of Capitalism*.

7 Topik, "The Integration of the World Coffee Market," p. 37.

8 Mario K. Samper, "Café, Trabajo y Sociedad en Centroamérica (1870–1930): Una Historia Común y Divergente," in Víctor Hugo Acuña Ortega (ed.), *Las Repúblicas Agroexportadoras. Tomo IV: Historía General de Centroamérica* (Costa Rica: FLACSO – Programa Costa Rica, 1994). Rural smallholders who provide some of their subsistence needs through home production but must ultimately sell their labor power to survive are often referred to as "semi-proletarian." See Carmen Diana Deere and Alain de Janvry, "A Conceptual Framework for the Empirical Analysis of Peasants," *American Journal of Agricultural Economics* 61 (1979).

9 Topik, "Coffee," pp. 61–7.

10 See United Nations Conference on Trade and Development and International Institute for Trade and Development (IISD), *Sustainability in the Coffee Sector: Exploring Opportunities for International Cooperation* (Geneva: UNCTAD; Winnipeg: IISD, 2003), pp. 5–6.

11 UNCTAD and IISD, *Sustainability in the Coffee Sector*, p. 6 n. 16.

12 This brief account of Costa Rican history relies in particular on the work of Winson, *Coffee and Democracy in Modern Costa Rica*; Williams, *States and Social Evolution*; Paige, *Coffee and Power*.

13 Cathy Payne, "Heavy Coffee Consumption Linked to Higher Death Risk," *USA Today* (August 16, 2013), http://www.usatoday.com/story/news/nation/2013/08/15/coffee-consumption-death-risk/2655855, accessed November 19, 2013.

14 Pendergrast, *Uncommon Grounds*, pp. 229–62.
15 Pendergrast, *Uncommon Grounds*, p. 280.
16 This understanding of international politics and power, along with the quote of "skewed," comes from Robert Cox, "Labour and Hegemony," *International Organization* 31: 3 (1977), p. 387. The reference to a "tug of war" is drawn from Wolf, *Europe and the People without History*, p. 148.

CHAPTER 3

1 "Social regulation" is drawn from the idea of "social efficiency" advanced by Lefeber and Vietorisz, "Meaning of Social Efficiency," as well as the ideas in Kevin Watkins, *Growth with Equity Is Good for the Poor* (Oxford: Oxfam GB, 2000); Michael A. Lebowitz, *Build It Now: Socialism for the Twenty-First Century* (New York: Monthly Review Press, 2006); Ananya Mukherjee Reed, *Human Development and Social Power: Perspectives from South Asia* (London: Routledge, 2008). The concept is more fully developed in Gavin Fridell, *Alternative Trade: Legacies for the Future* (Black Point, NS: Fernwood, 2013).
2 The general narrative and statistics on coffee prices, consumption, and exports in the decades leading up to the signing of the ICA draw in particular on Pendergrast, *Uncommon Grounds*.
3 Talbot, *Grounds for Agreement*.
4 Pendergrast, *Uncommon Grounds*, pp. 48–9.
5 Talbot, *Grounds for Agreement*; Gibbon and Ponte, *Trading Down*; Ellen Pay, *The Market for Organic and Fair-Trade Coffee: Study Prepared in the Framework of Fao Project GCP/RAF/404/GER, "Increasing Incomes and Food Security of Small Farmers in West and Central Africa through Exports of Organic and Fair-Trade Tropical Products"* (Rome: Food and Agriculture Organization of the United Nations, Trade and Markets Division, 2009); Inter-African Coffee Organization, *Improving African Coffee Processing and Market Access* (Abidjan: IACO, 2010).
6 Pendergrast, *Uncommon Grounds*, pp. 273–7.
7 Talbot, *Grounds for Agreement*, p. 59.
8 Oxfam International, *Mugged*.
9 Talbot, *Grounds for Agreement*, p. 111.

10 Oxfam International, *Mugged*; Inter-African Coffee Organization, *Improving African Coffee Processing and Market Access.*

11 According to Oxfam International, *Mugged*, p. 2, the same year that the ICO announced $45.2 million in funding for its projects, each of the world's four largest coffee companies (Kraft Foods Group, Nestlé, Sara Lee, and Procter & Gamble) had coffee brands worth $1 billion or more in annual sales.

12 ICO agreements are available at the ICO website (http://www.ico.org).

13 Anna Edgerton and Mario Sergio Lima, "Brazil to Buy Coffee above Market Price through Option Contracts," *Bloomberg.com* (August 7, 2013). Wexler, Lewis, and Josephs, "Brazil Drought Jolts Commodities' Prices."

14 Lefeber and Vietorisz, "Meaning of Social Efficiency," pp. 141, 158, 161.

15 Fridell, *Fair Trade Coffee.*

16 In one particular case in the mid-1990s, the use of "endosulfan" as an insecticide in Colombia was linked to more than 200 poisonings. See UNCTAD and IISD, *Sustainability in the Coffee Sector*; Robert A. Rice, "A Rich Brew from the Shade," *Americas* 50: 2 (1998); Dicum and Luttinger, *Coffee Book.*

CHAPTER 4

1 Daniele Giovannucci, Bryan Lewin, Rob Swinkels, and Panos Varangis, *Vietnam Coffee Sector Report* (Washington, DC: World Bank, 2004), p. ix.

2 Giovannucci et al., *Vietnam Coffee Sector Report*, p. 7.

3 The history of the Vietnamese coffee industry is drawn from Giovannucci et al., *Vietnam Coffee Sector Report*, as well as D. D'haeze, J. Deckers, D. Raes, T. A. Phong, and H. V. Loi, "Environmental and Socio-Economic Impacts of Institutional Reforms on the Agricultural Sector of Vietnam Land Suitability Assessment for Robusta Coffee in the Dak Gan Region," *Agriculture, Ecosystems and Environment* 105 (2005); Dang Thanh Ha and Gerald Shively, "Coffee Boom, Coffee Bust and Smallholder Response in Vietnam's Central Highlands," *Review of Development Economics* 12: 2 (2008); Larsen et al., "Global–Local Interactions." See also Gavin Fridell, "Coffee Statecraft:

Rethinking the Global Coffee Crisis, 1998–2002," *New Political Economy* (2013), accessed January 20, 2014.

4 Wood, *Empire of Capital*.

5 Giovannucci et al., *Vietnam Coffee Sector Report*, p. xi.

6 After 20 years of intense negotiations, Russia became a member of the WTO in 2012, an event much desired by coffee sectors in Brazil and Africa that have sought greater access to the Russian coffee market, the seventh largest in the world and the largest for instant coffee. See Adam Robert Green, "Russia–Africa Trade: Set for a WTO Boost?" *This is Africa* (November 22, 2012). States have also used regional free trade agreements to settle disputes over various barriers blocking the smooth transport of beans. See the case of the dispute between Tanzania and Rwanda, in Edward Ojulu, "Rwanda Squeezes Tanzania to Remove Barriers to Trade," *Bridges Africa* 1: 4 (October 25, 2012). For more on Vietnam's reluctance to take part in "sustainable" coffee certification, see Veronique Mistiaen, "A Better Future Is Percolating for Vietnam's Coffee," *Guardian* (March 26, 2012), http://www.guardian.co.uk/global-development/poverty-matters/2012/mar/26/better-future-vietnam-coffee-growth, accessed August 7, 2012.

7 Gowan, *The Global Gamble*.

8 Peter Baker, "What the Papers Say," *Coffee and Cocoa International* 40: 5 (November 2013), p. 44.

9 Neilson and Pritchard, *Value Chain Struggles*, pp. 107–29.

10 Giovannucci et al., *Vietnam Coffee Sector Report*.

11 In 2011, China was only the 12th-largest export market for Vietnamese coffee (in overall quantity), with the United States first, followed by European countries. However, as Vietnam provided 75 percent of China's coffee imports, it is poised to benefit from the anticipated rapid growth of the Chinese coffee market. Data from World Integrated Trade Solution (WITS), United Nations Commodity Trade Statistics Database (UN Comtrade), accessed November 18, 2013. See also Mistiaen, "Better Future."

12 Reuters, "Crippling Debts, Bankruptcies Brew Vietnam Coffee Crisis," *Voice of America* (August 14, 2013), http://www.voanews.com/content/reu-debt-bankruptcies-vietnam-coffee-crisis/1730013.html, accessed January 8, 2014.

13 Giovannucci et al., *Vietnam Coffee Sector Report*; D'haeze et al., "Environmental and Socio-Economic Impacts."

14 For an exception, see Mistiaen, "Better Future."
15 Giovannucci et al., *Vietnam Coffee Sector Report*, p. 67.
16 This is based on Vietnam's GDP (current US$), which ranked
    55th in 2011 according to World Bank data online (http://data.
    worldbank.org).

CHAPTER 5

1 Harvey, *New Imperialism*; McNally, *Another World*.
2 The calculation on big business advertising is taken from
   Dawson, *Consumer Trap*. The estimated cost of ending world
   hunger varies within the range of $30 billion to $50 billion. In
   2009, the Food and Agriculture Organization of the United
   Nations (FAO) estimated world governments needed to invest
   an additional $44 billion to end world hunger. See Jeffrey
   Donovan, "Ending World Hunger Will Require $44 Billion a
   Year (Correct)," *Bloomberg.com* (November 11, 2008), http://www.
   bloomberg.com/apps/news?pid=newsarchive&sid=axB4c8xDmB
   Ak, accessed November 26, 2012.
3 Anthony Winson, *The Industrial Diet: The Degradation of Food and
   the Struggle for Healthy Eating* (Vancouver: UBC Press, 2013).
4 Daviron and Ponte, *Coffee Paradox*.
5 Pendergrast, *Uncommon Grounds*, p. 202.
6 Steve Penfold, *The Donut: A Canadian History* (Toronto:
   University of Toronto Press, 2008), p. 171.
7 Patricia Cormack, "'True Stories' of Canada: Tim Hortons and
   the Branding of National Identity," *Cultural Sociology* 2: 3 (2008);
   Ari Alstedter, "Tim Hortons Cracks Top Five U.S. Coffee Shops
   in Zagat," *Bloomberg.com* (September 28, 2012), http://www.
   bloomberg.com/news/2012–09–28/tim-hortons-cracks-top-five-
   u-s-coffee-shops-in-zagat-su.html, accessed February 6, 2013.
8 Pendergrast, *Uncommon Grounds*, pp. 138–42; Heather Fowler-
   Salamini, "Women Coffee Sorters Confront the Mill Owners and
   the Veracruz Revolutionary State, 1915–1918," *Journal of Women's
   History* 14: 1 (2002).
9 The precarious situation of service sector coffee workers in
   Canada was conveyed to me in a telephone interview with
   Alex Dagg, director of Unite Here, Ontario Council, and
   Canadian director of Unite Here, Canada (November 7, 2006),

Toronto, Canada. See also Andrew Jackson, *"Good Jobs in Good Workplaces": Reflections on Medium-Term Labour Market Challenges* (Ottawa: Caledon Institute of Social Policy, 2003).

10 Starbucks, *Fiscal 2013 Annual Report* (Starbucks, 2013).

11 Mark Pendergrast, "The Starbucks Experience Going Global," *Tea and Coffee Trade Online* 176: 2 (2002); Jackson, *"Good Jobs."*

12 For more on CAW 3000, see Fridell, *Fair Trade Coffee*, pp. 252–6. For tax avoidance by Starbucks, see Peter Campbell and Martin Robinson, "Starbucks Doesn't Pay a Bean in UK Tax," *MailOnline* (October 15, 2012), http://www.dailymail.co.uk/news/article-2218192/Starbucks-tax-Coffee-chain-shortchanges-British-taxpayers-paying-just-8-6m-past-14-years.html, accessed February 18, 2013; Simon Neville and Jill Treanor, "Starbucks to Pay £20m in Tax Over Next Two Years After Customer Revolt," *Guardian* (December 6, 2012), http://www.guardian.co.uk/business/2012/dec/06/starbucks-to-pay-10m-corporation-tax, accessed February 18, 2013. For the Starbucks CEO promoting austerity, see Leslie Patton, "Starbucks' Schultz Urges Fellow CEOs to Halt Campaign Giving," *Bloomberg.com* (August 15, 2011), http://www.bloomberg.com/news/2011–08–15/starbucks-schultz-urges-fellow-ceos-to-boycott-campaign-giving.html, accessed February 18, 2013; Josh Eidelson, "Starbucks Tycoon Bullies the Baristas," *Nation* (January 30, 2013), http://www.thenation.com/article/172547/starbucks-tycoon-bullies-baristas, accessed February 18, 2013.

13 Marie-Christine Renard, "The Interstices of Globalization: The Example of Fair Coffee," *Sociologia Ruralis* 39: 4 (1999), p. 485.

14 Bacon, "Confronting the Coffee Crisis"; Jaffee, *Brewing Justice*. Much of the critical discussion on fair trade is drawn from Fridell, *Fair Trade Coffee*. See also Sarah Lyon and Mark Moberg (eds.), *Fair Trade and Social Justice: Global Ethnographies* (New York: New York University Press, 2010); Bradford L. Barham, Mercedez Callenes, Seth Gitter, Jessa Lewis, and Jeremy Weber, "Fair Trade/Organic Coffee, Rural Livelihoods, and the 'Agrarian Question': Southern Mexican Coffee Families in Transition," *World Development* 39: 1 (2011); Hudson et al., *Fair Trade, Sustainability and Social Change*.

15 The price comparison here is based on the conventional export price for Brazilian Arabicas and the fair trade minimum price. Both prices are "free on board" (FOB), which includes the cost

of transporting the beans to the export port. Fair trade farmers typically receive more of the FOB price than conventional farmers, although precisely how much varies depending on many local factors. Thanks to Bill Barrett of Planet Bean Coffee for pointing this out to the author in January 2014. Despite this difference, for many years conventional Brazilian Arabicas were well above the fair trade minimum price, making up for the difference and suggesting that today's fair trade price would have been fairly standard or low for most coffee farmers from 1976 to 1989.

16 Henry J. Frundt, *Fair Bananas: Farmers, Workers, and Consumers Strive to Change an Industry* (Tucson: University of Arizona Press, 2009).

17 For more on Planet Bean, see Fridell, *Fair Trade Coffee*, pp. 225–75. For more on JustUs! Coffee, see Stacey Byrne and Errol Sharpe, *In Pursuit of Justice: JustUs! Coffee Roasters Co-op and the Fair Trade Movement* (Black Point, NS: Fernwood, forthcoming).

18 See Gavin Fridell, "The Co-Operative and the Corporation: Competing Visions of the Future of Fair Trade," *Journal of Business Ethics* 86 (2009); Bambi Semroc, Elizabeth Baer, Joanne Sonenshine, and Marielle Canter Weikel, *Assessment of the Starbucks Coffee and Farmer Equity (C.A.F.E.) Practices Program FY08–FY10* (n.p.: Starbucks and Conservation International, 2012).

19 See Anton Foek, "Trademarking Coffee: Starbucks Cuts Ethiopia Deal," *CorpWatch* (May 8, 2007); Oxfam Australia, "Oxfam Celebrates Win-Win Outcome for Ethiopian Coffee Farmers and Starbucks," http://www.oxfam.org/en/node/174, accessed March 5, 2014. The total compensation of Howard Schultz in 2011 included bonus and exercised stock options; see Forbes.com, *CEO Compensation Rankings* (2011), http://www.forbes.com/lists/2011/12/ceo-compensation-11_Howard-D-Schultz_S9CA.html, accessed February 19, 2013. The GDP per capita (current US$) in Ethiopia in 2011 was $357; see World Bank data online (http://data.worldbank.org). This estimate was made by dividing Schultz's total pay in 2011 by the Ethiopian GDP per capita.

CHAPTER 6

1 Interview with Mauricio Galindo, head of operations, ICO (March 8, 2013), London, UK. For an important discussion on how

Brazil's original success in valorizing its coffee led to consensus among state officials over the benefits of price regulation, see Talbot, *Grounds for Agreement*.

2 For general discussion on the cost-price squeeze, see Tony Weis, *The Global Food Economy: The Battle for the Future of Farming* (Halifax, NS: Fernwood, 2007), pp. 70–8. For specifics on the Colombian crisis, see Diana Delgado, "Colombia Hikes Coffee Subsidies, Calls for Farmers' Strike to End," *Globe and Mail* (March 2, 2013); Michael Sheridan, "After the Colombian Coffee Strike: What Is $444 Million Really Worth?" *Daily Coffee News* (March 20, 2013).

3 The discussion on the "financialization of coffee" is taken from Jennifer Clapp's wider discussion on the "financialization of food." See Clapp, *Food* (Cambridge: Polity, 2011), pp. 125–57. For the data presented, see Marcelo Justo, "Los mercaderos detrás del aumento de precious de los alimentos," *BBC Mundo* (October 16, 2012), http://www.bbc.co.uk/mundo/noticias/2012/10/121010_alimentos_mercaderes_especulacion_precios_marcelo_jmp.shtml, accessed March 5, 2014. For a longer historical view on speculation and coffee, see Pendergrast, *Uncommon Grounds*.

4 Gowan, *Global Gamble*.

5 Rice, "Rich Brew from the Shade"; UNCTAD and IISD, *Sustainability in the Coffee Sector*; Dicum and Luttinger, *Coffee Book*. For a wider discussion on the Green Revolution and food, see Haroon Akram-Lodhi, *Hungry for Change: Farmers, Food Justice and the Agrarian Question* (Black Point, NS: Fernwood, 2013).

6 For a summary of over 50 reports on shade-grown versus full-sun, see Robert A. Rice, *The Ecological Benefits of Shade-Grown Coffee: The Case for Going Bird Friendly* (Washington, DC: Smithsonian Migratory Bird Center – National Zoological Park, 2010). See also Patricia Moguel and Victor M. Toledo, "Biodiversity Conservation in Traditional Coffee Systems of Mexico," *Conservation Biology* 13: 1 (1999); José Sarukhán and Jorge Larson, "When the Commons Become Less Tragic: Land Tenure, Social Organization, and Fair Trade in Mexico," in Joanna Burger, Elinor Ostrom, Richard B. Norgaard, David Policansky, and Bernard D. Goldstein (eds.), *Protecting the Commons: A Framework for Resource Management in the Americas* (Washington, DC: Island Press, 2001). Information also attained from personal e-mail communication with Rex

Weyler, co-founder of Greenpeace International (September 10, 2013).

7 Pay, *Market for Organic and Fair-Trade Coffee*, pp. 9–10.

8 For the impact of climate change on wild Arabicas see Jeff Green, "Coffee Beans Burn Towards Extinction," *Toronto Star* (November 12, 2012). For its impact on Central America in general, see Michael Sheridan, "356. Coffee Rust: An Inconvenient Truth," *CRS Coffeelands Blog* (May 6, 2013).

9 Miho Shirotori and Ana Cristina Molina, *South–South Trade: The Reality Check* (Geneva: UNCTAD, 2009); UNDP, *Human Development Report 2013: The Rise of the South – Human Progress in a Diverse World* (New York: UNDP, 2013).

10 For statistics on emerging coffee markets, see Coffee & Cocoa International, "Soluble Gains Market Share as Emerging Markets Evolve," *Coffee & Cocoa International* 40: 5 (November 2013), pp. 36–7. For new coffee trends in Brazil and Latin America, see Mario K. Samper and Steven C. Topik (eds.), *Crisis y Transformaciones del Mundo del Café: Dinámicas Locales y Estrategias Nacionales en un Periodo de Adversidad e Incertidumbre* (Bogotá: Editorial Pontificia Universidad Javeriana, 2012); Roberto Samora, "Update 2: Brazil Brews More Coffee as Beans Get Tastier – ABIC" (January 25, 2012), http://uk.reuters.com/article/2012/01/25/idUKL2E8CO15520120125, accessed August 14, 2012. See also Carlos Henrique Jorge Brando, "How to Increase Internal Consumption of Coffee: Brazil's Success Story" (n.p.: P&A Marketing International, 2008), http://www.slideshare.net/lemeph/how-to-increase-internal-coffee-consumption-presentation, accessed August 14, 2012.

11 Daniel Allen, "China's New Brew," *Asia Times* (March 11, 2011), http://www.atimes.com/atimes/China_Business/MC11Cb02.html, accessed August 14, 2012.

12 "Statement by the Chairperson of the InterAfrican Coffee Organisation (IACO) to the 110th Session of the International Coffee Council," read out by M. Aly Touré, representative of Côte d'Ivoire, on behalf of the IACO at the 110th Session of the ICO, London (March 7, 2013), http://dev.ico.org/documents/cy2012–13/wsiteenglish/council-12-e.htm, document ICC-110–8, accessed May 16, 2013.

13 Adolph A. Kumburu, director general, Tanzania Coffee Board, "Introducing: Tanzania Coffee Industry Development Strategy

(2011–2021)," presented to the 110th Session of the ICO, London (March 7, 2013), http://www.ico.org/presents/1213/march-tanzania.pdf, accessed May 16, 2013.

14 See Daviron and Ponte, *Coffee Paradox*. The discussion here on Ethiopian coffee is based on Oxfam Australia, "Oxfam Celebrates Win-Win Outcome for Ethiopian Coffee Farmers and Starbucks"; Overseas Development Institute (ODI), *Ethiopia Trademarking and Licensing Initiative: Supporting a Better Deal for Coffee Farmers through Aid for Trade* (London: ODI, 2009); ICO, *Ethiopia: Country Datasheet* (London: ICO, 2011); Coffee & Cocoa International, "ECX and Outgoing CEO Praised as Specialty Issues Begin to Be Addressed," *Coffee & Cocoa International* (January 2013).

15 See Light Years IP website, http://www.lightyearsip.net/projects/ethiopiancoffee, accessed May 22, 2013.

16 The discussion of CLAC is based on Marco Coscione, *CLAC and the Defense of the Small Producer*, trans. Lizzy Solano Guzmán (Black Point, NS: Fernwood, 2014), English translation of *La CLAC y la defensa del pequeño productor* (Santo Domingo: CLAC and Editorial Funglode, 2012).

17 For more on the coffee crisis, see ICO, *Report on the Outbreak of Coffee Leaf Rust in Central America* (London: ICO, May 13, 2013); Gavin Fridell, "Coffee Crisis in Central America," *Watershed Sentinel* 23: 4 (September/October 2013). Ecologist John Vandermeer has argued that monocrop cultivation has led to the decline of white halo fungus, which tends to restrain the spread of coffee leaf rust, and may have played a significant role in the outbreak. See Emma Bryce, "Fighting Off the Coffee Curse," *New York Times* (February 8, 2013).

18 Anna Edgerton and Mario Sergio Lima, "Brazil to Buy Coffee above Market Price through Option Contracts," *Bloomberg.com* (August 7, 2013).

19 Ilan Kapoor, "Participatory Development, Complicity and Desire," *Third World Quarterly* 26: 8 (2005), p. 1218.

Selected readings

Given its popularity and importance, numerous works have been written on coffee over the years, including excellent global assessments that are drawn on throughout this book: Robert H. Bates, *Open-Economy Politics: The Political Economy of the World Coffee Trade* (Princeton, NJ: Princeton University Press, 1997); Steven C. Topik, "Coffee," in Steven C. Topik and Allen Wells (eds.), *The Second Conquest of Latin America: Coffee, Henequen, and Oil During the Export Boom* (Austin: University of Texas Press, 1998); John M. Talbot, *Grounds for Agreement: The Political Economy of the Coffee Commodity Chain* (Oxford: Rowman & Littlefield, 2004); Benoit Daviron and Stefano Ponte, *The Coffee Paradox: Global Markets, Commodity Trade and the Elusive Promise of Development* (London: Zed Books, 2005). Out of necessity, this short annotated list of selected reading focuses on English-language materials, but many important works have been written in other languages, in particular Spanish. An excellent place to start is Mario K. Samper and Steven C. Topik (eds.), *Crisis y Transformaciones del Mundo del Café: Dinámicas Locales y Estrategias Nacionales en un Periodo de Adversidad e Incertidumbre* (Bogotá: Editorial Pontificia Universidad Javeriana, 2012).

There are countless journalistic works on coffee, among the best of which are Mark Pendergrast, *Uncommon Grounds: The History of Coffee and How It Transformed Our World*, 2nd edn. (New York: Basic Books, 2010), which provides rich empirical and narrative details drawn on throughout this book; Anthony

Wild, *Coffee: A Dark History* (New York: W. W. Norton, 2005); Gregory Dicum and Nina Luttinger, *The Coffee Book: Anatomy of an Industry from Crop to the Last Drop*, rev. edn. (New York: New Press, 2006). The UNCTAD statistical database (http://unctadstat.unctad.org) provides core historical price data here; the ICO offers invaluable daily price data and analysis (http://www.ico.org); and WITS (https://wits.worldbank.org/WITS), combined with the United Nations Commodity Trade Statistics Database (UN-COMTRADE) (http://comtrade.un.org), provide invaluable data on imports and exports. An excellent "insiders" magazine is *Coffee & Cocoa International* (http://www.coffeeandcocoa.net).

Chapter 1 draws on the sources just listed, along with other works dealing with global value chains, including Niels Fold and Bill Pritchard, *Cross-Continental Food Chains* (London: Routledge, 2005); Henry Bernstein and Liam Campling, "Commodity Studies and Commodity Fetishism I: *Trading Down*," *Journal of Agrarian Change* 6: 2 (2006); "Commodity Studies and Commodity Fetishism II: 'Profits with Principles'?," *Journal of Agrarian Change* 6: 3 (2006); Jennifer Bair (ed.), *Frontiers in Commodity Chain Research* (Stanford, CA: Stanford University Press, 2009); Jeffrey Neilson and Bill Pritchard, *Value Chain Struggles: Institutions and Governance in the Plantation Districts of South India* (Oxford: Wiley-Blackwell, 2009); and excellent reports on price volatility and corporate power in the coffee chain by Oxfam International, *Mugged: Poverty in Your Coffee Cup* (Boston and Washington, DC: Oxfam International, 2002); Duncan Green, "Conspiracy of Silence: Old and New Directions on Commodities," paper presented at the Strategic Dialogue on Commodities, Trade, Poverty and Sustainable Development Conference, Faculty of Law, Barcelona (June 13–15, 2005).

The concept of "coffee statecraft" draws on the political economy works of Peter Gowan, *The Global Gamble:*

Washington's Faustian Bid for World Dominance (London: Verso, 1999); David Harvey, *The New Imperialism* (Oxford: Oxford University Press, 2003); Ellen Meiksins Wood, *Empire of Capital* (London: Verso, 2005); Giovanni Arrighi, *Adam Smith in Beijing: Lineages of the Twenty-First Century* (London: Verso, 2007). The critique of free trade is based on Ha-Joon Chang, *Bad Samaritans: The Myth of Free Trade and the Secret History of Capitalism* (New York: Bloomsbury Press, 2008); Dimitris Milonakis and Ben Fine, *From Political Economy to Economics: Method, the Social and the Historical in the Evolution of Economic Theory* (London: Routledge, 2009); John Quiggin, *Zombie Economics: How Dead Ideas Still Walk among Us* (Princeton, NJ: Princeton University Press, 2010); as well as the idea of the "free trade fantasy" advanced by Jodi Dean, *Democracy and Other Neoliberal Fantasies: Communicative Capitalism and Left Politics* (Durham, NC: Duke University Press, 2009).

Chapter 2 draws on Topik, "Coffee," and Pendergrast, *Uncommon Grounds*, as well as highly significant works on coffee and colonial and post-independence state building, including Liisa L. North, *Bitter Grounds: Roots of Revolt in El Salvador*, 2nd edn. (Westport, CT: Lawrence Hill, 1985); Jim Handy, *Revolution in the Countryside: Rural Conflict and Agrarian Reform in Guatemala, 1944–1954* (Chapel Hill: University of North Carolina Press, 1994); David McCreery, *Rural Guatemala, 1760–1940* (Stanford, CA: Stanford University Press, 1994); Robert G. Williams, *States and Social Evolution: Coffee and the Rise of National Governments in Central America* (Chapel Hill: University of North Carolina Press, 1994); William Gervase Clarence-Smith and Steven Topik (eds.), *The Global Coffee Economy in Africa, Asia, and Latin America, 1500–1989* (New York: Cambridge University Press, 2003). The discussion of Costa Rica draws on John A. Booth, "Costa Rica: The Roots of Democratic Stability," in

Larry Diamond, Jonathan Hartlyn, Juan J. Linz, and Seymour Martin Lipset (eds.), *Democracy in Developing Countries: Latin America* (Boulder: Lynne Rienner, 1988); Anthony Winson, *Coffee and Democracy in Modern Costa Rica* (Toronto: Between the Lines, 1989); Jeffery M. Paige, *Coffee and Power: Revolution and the Rise of Democracy in Central America* (Cambridge, MA: Harvard University Press, 1997); Marc Edelman, *Peasants against Globalization: Rural Social Movements in Costa Rica* (Stanford, CA: Stanford University Press, 1999); and Williams, *States and Social Evolution.*

For the wider discussion on the history of the world capitalist system, see Karl Polanyi, *The Great Transformation: The Political and Economic Origins of Our Time* (Boston: Beacon Press, 1944); Robert Brenner, "The Origins of Capitalist Development: A Critique of Neo-Smithian Marxism," *New Left Review* 104 (1977); Sidney W. Mintz, *Sweetness and Power: The Place of Sugar in Modern History* (New York: Penguin, 1985); Eric R. Wolf, *Europe and the People without History* (Berkeley: University of California Press, 1997); Ellen Meiksins Wood, *The Origin of Capitalism* (New York: Monthly Review Press, 1999).

The works of Bates, *Open-Economy Politics*, and Talbot, *Grounds for Agreement*, are central to chapter 3. The idea of "social regulation" is drawn from Louis Lefeber and Thomas Vietorisz, "The Meaning of Social Efficiency," *Review of Political Economy* 19: 2 (2007), as well as Kevin Watkins, *Growth with Equity Is Good for the Poor* (Oxford: Oxfam GB, 2000); Michael A. Lebowitz, *Build It Now: Socialism for the Twenty-First Century* (New York: Monthly Review Press, 2006); Ananya Mukherjee Reed, *Human Development and Social Power: Perspectives from South Asia* (London: Routledge, 2008). This concept is more fully developed in Gavin Fridell, *Alternative Trade: Legacies for the Future* (Black Point, NS: Fernwood, 2013). For more on international commodity agreements, see Michael Barratt Brown, *Fair Trade: Reform and Realities in the International Trading System* (London: Zed Books, 1993); Peter Robbins, *Stolen Fruit: The Tropical Commodities Disaster* (London: Zed Books, 2003); Thomas Lines, *Making Poverty: A History* (London: Zed Books, 2008).

Chapter 4 draws social and economic data on Vietnamese coffee from the World Bank report, Daniele Giovannucci, Bryan Lewin, Rob Swinkels, and Panos Varangis, *Vietnam Coffee Sector Report* (Washington, DC: World Bank, 2004). Other excellent accounts include D. D'haeze, J. Deckers, D. Raes, T. A. Phong, and H. V. Loi, "Environmental and Socio-Economic Impacts of Institutional Reforms on the Agricultural Sector of Vietnam Land Suitability Assessment for Robusta Coffee in the Dak Gan Region," *Agriculture, Ecosystems and Environment* 105(2005); Dang Thanh Ha and Gerald Shively, "Coffee Boom, Coffee Bust and Smallholder Response in Vietnam's Central Highlands," *Review of Development Economics* 12: 2 (2008); Jytte Agergaard Larsen, Niels Fold, and Katherine Gough, "Global–Local Interactions: Socioeconomic and Spatial Dynamics in Vietnam's Coffee Frontier," *Geographical Journal* 175: 2 (2009). The discussion on the Vietnam War and the post-war era is taken from Gabriel Kolko, *Anatomy of a War: Vietnam, the United States, and the Modern Historical Experience* (New York: New Press, 1994).

Chapter 5 draws on a wide-ranging literature on corporate power and social responsibility, including works by Michael Dawson, *The Consumer Trap: Big Business Marketing in American Life* (Urbana: University of Illinois Press, 2003); Susanne Soederberg, *Corporate Power and Ownership in Contemporary Capitalism: The Politics of Resistance and Domination* (London: Routledge, 2010); Anthony Winson, *The Industrial Diet: The Degradation of Food and the Struggle for Healthy Eating* (Vancouver: UBC Press, 2013), as well as Pendergrast, *Uncommon Grounds*. For more on Tim Hortons, see Patricia Cormack, "'True Stories' of Canada: Tim Hortons and the Branding of National Identity," *Cultural Sociology* 2: 3 (2008); Steve Penfold, *The Donut: A Canadian History* (Toronto: University of Toronto Press, 2008); and for Starbucks see Stefano Ponte, "The 'Latte Revolution'? Regulation, Markets and Consumption in the Global Coffee Chain," *World Development* 30: 7 (2002); Gavin Fridell, "The Co-Operative and the Corporation: Competing Visions of the Future of Fair Trade," *Journal of Business Ethics* 86 (2009).

A great deal has been written on fair trade in recent years, including Dean Cycon, *Javatrekker: Dispatches from the World of Fair Trade Coffee* (White River Junction, VT: Chelsea Green, 2007); Gavin Fridell, *Fair Trade Coffee: The Prospects and Pitfalls of Market-Driven Social Justice* (Toronto: University of Toronto Press, 2007); Daniel Jaffee, *Brewing Justice: Fair Trade Coffee, Sustainability, and Survival* (Berkeley: University of California Press, 2007); Ian Hudson, Mark Hudson, and Mara Fridell, *Fair Trade, Sustainability and Social Change* (New York: Palgrave Macmillan, 2013). Excellent edited volumes with case studies include Laura T. Raynolds, Douglas L. Murray and John Wilkinson (eds.), *Fair Trade: The Challenges of Transforming Globalization* (London: Routledge, 2007); Christopher M. Bacon, V. Ernesto Méndez, Stephen R. Gliessman, David Goodman, and Jonathan A. Fox (eds.), *Confronting the Coffee Crisis: Fair Trade, Sustainable Livelihoods and Ecosystems in Mexico and Central America* (Cambridge, MA: MIT Press, 2008); Sarah Lyon and Mark Moberg (eds.), *Fair Trade and Social Justice: Global Ethnographies* (New York: New York University Press, 2010). Some important articles include Marie-Christine Renard, "The Interstices of Globalization: The Example of Fair Coffee," *Sociologia Ruralis* 39: 4 (1999); Mark S. LeClair, "Fighting the Tide: Alternative Trade Organizations in the Era of Global Free Trade," *World*

Development 30: 6 (2002); Christopher Bacon, "Confronting the Coffee Crisis: Can Fair Trade, Organic, and Specialty Coffees Reduce Small-Scale Farmer Vulnerability in Northern Nicaragua?" *World Development* 33: 3 (2005); Darryl Reed, "What Do Corporations Have to Do with Fair Trade? Positive and Normative Analysis from a Value Chain Perspective," *Journal of Business Ethics* 86: 1 (2009). Two recent excellent ethnographies are Sarah Lyon, *Coffee and Community: Maya Farmers and Fair-Trade Markets* (Boulder: University Press of Colorado, 2010); Paige West, *From Modern Production to Imagined Primitive: The Social World of Coffee from Papua New Guinea* (Durham, NC: Duke University Press, 2012). For more on organic coffee, see Tad Mutersbaugh, "The Number Is the Beast: A Political Economy of Organic-Coffee Certification and Producer Unionism," *Environment and Planning A* 34 (2002). For a discussion of fair trade North, see Stacey Byrne and Errol Sharpe, *In Pursuit of Justice: Just Us! Coffee Roasters Co-op and the Fair Trade Movement* (Black Point, NS: Fernwood, forthcoming).

The final chapter draws on the political economy of food and food sovereignty, including Tony Weis, *The Global Food Economy: The Battle for the Future of Farming* (Halifax, NS: Fernwood, 2007); Jennifer Clapp, *Food* (Cambridge: Polity, 2011); and Haroon Akram-Lodhi, *Hungry for Change: Farmers, Food Justice and the Agrarian Question* (Black Point, NS: Fernwood, 2013), which contains an excellent discussion on the connection between the state, powerful philanthropic organizations, and biotechnology applied to agriculture. For more on the developmental state, see Mark Weisbrot, Dean Baker, and David Rosnick, *The Scorecard on Development: 25 Years of Diminished Progress* (Washington, DC: Center for Economic and Policy Research, 2005); and Chang, *Bad Samaritans*. For more on the environmental sustainability of small-farmer, shade-grown coffee, see Robert A. Rice, "A

Rich Brew from the Shade," *Americas* 50: 2 (1998); Patricia Moguel and Victor M. Toledo, "Biodiversity Conservation in Traditional Coffee Systems of Mexico," *Conservation Biology* 13: 1 (1999); José Sarukhán and Jorge Larson, "When the Commons Become Less Tragic: Land Tenure, Social Organization, and Fair Trade in Mexico," in Joanna Burger, Elinor Ostrom, Richard B. Norgaard, David Policansky, and Bernard D. Goldstein (eds.), *Protecting the Commons: A Framework for Resource Management in the Americas* (Washington, DC: Island Press, 2001). On the "rise of the South," see UNDP, *Human Development Report 2013: The Rise of the South – Human Progress in a Diverse World* (New York: UNDP, 2013).

An excellent resource on fair trade South and CLAC is Marco Coscione, *CLAC and the Defense of the Small Producer*, trans. Lizzy Solano Guzmán (Black Point, NS: Fernwood, 2014), English translation of *La CLAC y la defensa del pequeño productor* (Santo Domingo: CLAC and Editorial Funglode, 2012). For more on coffee and global justice, see Global Exchange (www.globalexchange.org) and Oxfam International (www.oxfam.org).

Index